Lecture Notes in Business Information Processing 458

Series Editors

Wil van der Aalst
RWTH Aachen University, Aachen, Germany

John Mylopoulos
University of Trento, Trento, Italy

Sudha Ram
University of Arizona, Tucson, AZ, USA

Michael Rosemann
Queensland University of Technology, Brisbane, QLD, Australia

Clemens Szyperski
Microsoft Research, Redmond, WA, USA

More information about this series at https://link.springer.com/bookseries/7911

Claudio Di Ciccio · Remco Dijkman ·
Adela del Río Ortega ·
Stefanie Rinderle-Ma (Eds.)

Business Process Management Forum

BPM 2022 Forum
Münster, Germany, September 11–16, 2022
Proceedings

 Springer

Editors
Claudio Di Ciccio (iD)
Sapienza University of Rome
Rome, Italy

Remco Dijkman (iD)
Eindhoven University of Technology
Eindhoven, The Netherlands

Adela del Río Ortega (iD)
Universidad de Sevilla
Seville, Spain

Stefanie Rinderle-Ma (iD)
Technical University of Munich
Garching, Germany

ISSN 1865-1348   ISSN 1865-1356 (electronic)
Lecture Notes in Business Information Processing
ISBN 978-3-031-16170-4   ISBN 978-3-031-16171-1 (eBook)
https://doi.org/10.1007/978-3-031-16171-1

This Springer imprint is published by the registered company Springer Nature Switzerland AG
The registered company address is: Gewerbestrasse 11, 6330 Cham, Switzerland

Preface

This volume contains all papers presented at the BPM Forum of the 20th International Conference on Business Process Management (BPM 2022), held during September 11–16, 2022, in Münster, Germany. Similarly to previous years, the BPM Forum hosted innovative research contributions characterized by their high potential to stimulate interesting discussion and scientific debate, although not yet reaching the rigorous technical quality criteria required to be presented at the main conference. In this sense, the BPM Forum papers characterize themselves by novel ideas about emergent BPM topics. This year, the conference received a total of 114 submissions, out of which 97 entered the review phase. The review process for each paper involved at least three Program Committee members and one Senior Program Committee member. In the end, 23 papers were accepted, and 13 papers were invited to the BPM Forum (the latter being compiled in this volume).

BPM 2021 offered a hybrid attendance mode, allowing participants to attend the conference online or physically, thus providing the opportunity to connect again in person. BPM 2022 strived for a full in-person celebratory 20th anniversary conference, flanked by a multitude of events, such as the Blockchain, CEE, and RPA fora, workshops, tutorials, and wonderful social events, that provided the opportunity for networking and exchanging the latest research ideas. We would like to express our gratitude to Jörg Becker as the General Chair of BPM 2022, together with the Organizing Committee Chairs Katrin Bergener and Armin Stein and their group. The Münster team did an invaluable job in planning and organizing an unforgettable conference, especially in light of the still challenging times we are living in, with the high degree of uncertainty this also adds to the organizational tasks.

We also thank the members of the Program Committees and the external reviewers. They made a rigorous and extensive review procedure possible and thus enabled the high-quality research output reflected by the papers in both the main conference and BPM Forum proceedings. Finally, we acknowledge our sponsors for their support in making BPM 2022 happen: Celonis, SAP Signavio, and MR.KNOW as platinum sponsors; cronos, Provinzial, and viadee as bronze sponsors; and Deutsche Forschungsgemeinschaft (German Research Foundation, DFG), Springer, the University of Münster, and the European Research Center for Information Systems (ERCIS) as academic sponsors.

September 2022

Claudio Di Ciccio
Remco Dijkman
Adela del Río Ortega
Stefanie Rinderle-Ma

Organization

The 20th International Conference on Business Process Management (BPM 2022) was organized by the University of Münster, and took place in Münster, Germany.

Steering Committee

Mathias Weske (Chair)	HPI, University of Potsdam, Germany
Wil van der Aalst	RWTH Aachen University, Germany
Boualem Benatallah	University of New South Wales, Australia
Jörg Desel	Fernuniversität Hagen, Germany
Marlon Dumas	University of Tartu, Estonia
Jan Mendling	Humboldt Universität zu Berlin, Germany
Manfred Reichert	University of Ulm, Germany
Stefanie Rinderle-Ma	Technical University of Munich, Germany
Hajo Reijers	Utrecht University, The Netherlands
Michael Rosemann	Queensland University of Technology, Australia
Shazia Sadiq	University of Queensland, Australia
Barbara Weber	University of St. Gallen, Switzerland

Executive Committee

General Chair

Jörg Becker	University of Münster, Germany

Main Conference Program Committee Chairs

Claudio Di Ciccio (Track I Chair)	Sapienza University of Rome, Italy
Remco Dijkman (Track II Chair)	Eindhoven University of Technology, The Netherlands
Adela del Río Ortega (Track III Chair)	University of Seville, Spain
Stefanie Rinderle-Ma (Consolidation Chair)	Technical University of Munich, Germany

Workshop Chairs

Cristina Cabanillas	University of Seville, Spain
Agnes Koschmider	Kiel University, Germany
Niels F. Garmann-Johnsen	University of Agder, Norway

Demonstration and Resources Chairs

Christian Janiesch	TU Dortmund, Germany
Chiara Di Francescomarino	Fondazione Bruno Kessler, Italy
Thomas Grisold	University of Liechtenstein, Liechtenstein

Tutorial Chairs

Bettina Distel	University of Münster, Germany
Minseok Song	POSTECH, South Korea

Industry Forum Chairs

Jan vom Brocke	University of Liechtenstein, Liechtenstein
Jan Mendling	Humboldt-Universität zu Berlin, Germany
Michael Rosemann	Queensland University of Technology, Australia

Blockchain Forum Chairs

Raimundas Matulevicius	University of Tartu, Estonia
Qinghua Lu	CSIRO Data61, Australia
Walid Gaaloul	Télécom SudParis, France

RPA Forum Chairs

Andrea Marrella	Sapienza University of Rome, Italy
Bernhard Axmann	Technical University of Ingolstadt, Germany

Central and Eastern European Forum Chairs

Vesna Bosilj Vukšić	University of Zagreb, Croatia
Renata Gabryelczyk	University of Warsaw, Poland
Mojca Indihar Štemberger	University of Ljubljana, Slovenia
Andrea Kő	Corvinus University of Budapest, Hungary

Doctoral Consortium Chairs

Hajo Reijers	Utrecht University, The Netherlands
Robert Winter	University of St. Gallen, Switzerland

BPM Dissertation Award Chair

Jan Mendling	Humboldt Universität zu Berlin, Germany

Journal First Track Chairs

Matthias Weidlich	Humboldt Universität zu Berlin, Germany
Amy Van Looy	Ghent University, Belgium

Publicity Chairs

Flavia Maria Santoro   University of the State of Rio de Janeiro, Brazil
Abel Armas Cervantes   University of Melbourne, Australia
Manuel Resinas   University of Seville, Spain

Organizing Committee Chairs

Katrin Bergener   University of Münster, Germany
Armin Stein   University of Münster, Germany

Track I: Foundations

Senior Program Committee

Chiara Di Francescomarino   Fondazione Bruno Kessler, Italy
Dirk Fahland   Eindhoven University of Technology,
                                  The Netherlands
Chiara Ghidini   Fondazione Bruno Kessler, Italy
Thomas Hildebrandt   University of Copenhagen, Denmark
Richard Hull   New York University, USA
Sander J. J. Leemans   Queensland University of Technology, Australia
Andrea Marrella   Sapienza University of Rome, Italy
Fabrizio Maria Maggi   Free University of Bozen-Bolzano, Italy
Marco Montali   Free University of Bolzano-Bozen, Italy
Oscar Pastor Lopez   Universitat Politècnica de València, Spain
Artem Polyvyanyy   University of Melbourne, Australia
Manfred Reichert   University of Ulm, Germany
Arthur ter Hofstede   Queensland University of Technology, Australia
Wil van der Aalst   RWTH Aachen University, Germany
Jan Martijn van der Werf   Utrecht University, The Netherlands
Hagen Voelzer   IBM Research Europe, Germany
Matthias Weidlich   Humboldt-Universität zu Berlin, Germany
Mathias Weske   HPI, University of Potsdam, Germany

Program Committee

Lars Ackermann   University of Bayreuth, Germany
Adriano Augusto   University of Melbourne, Australia
Ahmed Awad   University of Tartu, Estonia
Patrick Delfmann   University of Koblenz-Landau, Germany
Rik Eshuis   Eindhoven University of Technology,
                                  The Netherlands
Peter Fettke   DFKI and Saarland University, Germany

Valeria Fionda	University of Calabria, Italy
Ulrich Frank	University of Duisburg-Essen, Germany
María Teresa Gómez-López	University of Seville, Spain
Guido Governatori	CSIRO Data61, Australia
Gianluigi Greco	University of Calabria, Italy
Giancarlo Guizzardi	Free University of Bozen-Bolzano, Italy
Antonella Guzzo	University of Calabria, Italy
Akhil Kumar	Pennsylvania State University, USA
Irina Lomazova	National Research University Higher School of Economics, Russia
Qinghua Lu	CSIRO Data61, Australia
Xixi Lu	Utrecht University, The Netherlands
Felix Mannhardt	Eindhoven University of Technology, The Netherlands
Werner Nutt	Free University of Bozen-Bolzano, Italy
Chun Ouyang	Queensland University of Technology, Australia
Luigi Pontieri	National Research Council of Italy (CNR), Italy
Daniel Ritter	SAP, Germany
Andrey Rivkin	Free University of Bozen-Bolzano, Italy
Arik Senderovich	University of Toronto, Canada
Tijs Slaats	University of Copenhagen, Denmark
Monique Snoeck	KU Leuven, Belgium
Ernest Teniente	Universitat Politècnica de Catalunya, Spain
Eric Verbeek	Eindhoven University of Technology, The Netherlands
Karsten Wolf	University of Rostock, Germany
Francesca Zerbato	University of St. Gallen, Switzerland

Track II: Engineering

Senior Program Committee

Boualem Benatallah	University of New South Wales, Australia
Andrea Burattin	Technical University of Denmark, Denmark
Josep Carmona	Universitat Politècnica de Catalunya, Spain
Jochen De Weerdt	Katholieke Universiteit Leuven, Belgium
Marlon Dumas	University of Tartu, Estonia
Avigdor Gal	Technion, Israel
Massimo Mecella	Sapienza University of Rome, Italy
Jorge Munoz-Gama	Pontificia Universidad Católica de Chile, Chile
Luise Pufahl	TU Berlin, Germany
Hajo A. Reijers	Utrecht University, The Netherlands
Shazia Sadiq	University of Queensland, Australia

Pnina Soffer	University of Haifa, Israel
Boudewijn van Dongen	Eindhoven University of Technology, The Netherlands
Barbara Weber	University of St. Gallen, Switzerland
Ingo Weber	TU Berlin, Germany
Moe Thandar Wynn	Queensland University of Technology, Australia

Program Committee

Marco Aiello	University of Stuttgart, Germany
Robert Andrews	Queensland University of Technology, Australia
Abel Armas Cervantes	University of Melbourne, Australia
Cristina Cabanillas	University of Seville, Spain
Fabio Casati	University of Trento, Italy
Massimiliano de Leoni	University of Padua, Italy
Johannes De Smedt	KU Leuven, Belgium
Benoît Depaire	Hasselt University, Belgium
Joerg Evermann	Memorial University of Newfoundland, Canada
Walid Gaaloul	Télécom SudParis, France
Luciano García-Bañuelos	Tecnológico de Monterrey, Mexico
Laura Genga	Eindhoven University of Technology, The Netherlands
Daniela Grigori	Université Paris Dauphine-PSL, France
Georg Grossmann	University of South Australia, Australia
Mieke Jans	Hasselt University, Belgium
Anna Kalenkova	University of Melbourne, Australia
Dimka Karastoyanova	University of Groningen, The Netherlands
Agnes Koschmider	Kiel University, Germany
Henrik Leopold	Kühne Logistics University, Germany
Francesco Leotta	Sapienza Università di Roma, Italy
Elisa Marengo	Free University of Bozen-Bolzano, Italy
Rabeb Mizouni	Khalifa University, United Arab Emirates
Timo Nolle	Technical University of Darmstadt, Germany
Helen Paik	University of New South Wales, Australia
Cesare Pautasso	University of Lugano, Switzerland
Pierluigi Plebani	Politecnico di Milano, Italy
Pascal Poizat	Université Paris Nanterre, France
Simon Poon	University of Sydney, Australia
Barbara Re	Università di Camerino, Italy
Manuel Resinas	University of Seville, Spain
Stefan Schönig	Universität Regensburg, Germany
Marcos Sepúlveda	Pontificia Universidad Católica de Chile, Chile

Track III: Management

Senior Program Committee

Program Committee

Kanika Goel	Queensland University of Technology, Australia
Tomislav Hernaus	University of Zagreb, Croatia
Christian Janiesch	TU Dresden, Germany
Andrea Kő	Corvinus University of Budapest, Hungary
John Krogstie	Norwegian University of Science and Technology, Norway
Michael Leyer	University of Rostock, Germany
Alexander Mädche	Karlsruhe Institute of Technology, Germany
Martin Matzner	FAU Erlangen-Nürnberg, Germany
Ralf Plattfaut	Fachhochschule Südwestfalen, Germany
Geert Poels	Ghent University, Belgium
Gregor Polancic	University of Maribor, Slovenia
Pascal Ravesteijn	HU University of Applied Sciences Utrecht, The Netherlands
Kate Revoredo	Wirtschaftsuniversität Wien, Austria
Dennis Riehle	University of Muenster, Germany
Stefan Sackmann	University of Halle-Wittenberg, Germany
Estefanía Serral	KU Leuven, Belgium
Oktay Turetken	Eindhoven University of Technology, The Netherlands
Inge van de Weerd	Utrecht University, The Netherlands
Irene Vanderfeesten	Open University of the Netherlands, The Netherlands
Axel Winkelmann	University of Wuerzburg, Germany
Bastian Wurm	Wirtschaftsuniversität Wien, Austria
Michael Zur Muehlen	Stevens Institute of Technology, USA

Additional Reviewers

Faria Khandaker	Manuel Weber
Vinicius Stein Dani	Hendrik Wache
Vladimir Bashkin	Wouter van der Waal
Christoph Drodt	Lukas-Valentin Herm
Boming Xia	Sandra Zilker
Yue Liu	Sven Weinzierl
Monika Kaczmarek-Heß	Mario Nadj
Nour Assy	Leonard Nake
Ebaa Alnazer	Ulrich Gnewuch
Dominik Janssen	Carolin Vollenberg
Robin Pesl	Iris Beerepoot
Brian Setz	Víctor Gálvez
Lorenzo Rossi	Laura Lohoff
Christoph Tomitza	Gregor Kipping

Florian Kragulj

Sebastian Dunzer

Johannes Damarowsky

Peyman Toreiniz

Martin Böhmer

Arjen Wierikx

Contents

Predictive Process Monitoring

Modeling and Design

Why Do Banks Find Business Process Compliance so Challenging? An Australian Perspective

Nigel Adams[⊠], Adriano Augusto, Michael Davern, and Marcello La Rosa

University of Melbourne, Melbourne, Australia
naadam@student.unimelb.edu.au,
{a.augusto,m.davern,m.larosa}@unimelb.edu.au

Abstract. Banks play an intrinsic role in any modern economy, recycling capital from savers to borrowers. They are heavily regulated and there have been a significant number of well publicized compliance failings in recent years. This is despite Business Process Compliance (BPC) being both a well researched domain in academia and one where significant progress has been made. This study seeks to determine why Australian banks find BPC so challenging. We interviewed 22 senior managers from a range of functions within the four major Australian banks to identify the key challenges. Not every process in every bank is facing the same issues, but in processes where a bank is particularly challenged to meet its compliance requirements, the same themes emerge. The compliance requirement load they bear is excessive, dynamic and complex. Fulfilling these requirements relies on impenetrable spaghetti processes, and the case for sustainable change remains elusive, locking banks into a fail-fix cycle that increases the underlying complexity. This paper proposes a conceptual framework that identifies and aggregates the challenges, and a circuit-breaker approach as an "off ramp" to the fail-fix cycle.

1 Introduction

Banks play an intrinsic role in any modern economy, they recycle capital between savers and borrowers and are tightly regulated. In the last five years, Australian regulators have highlighted multiple compliance issues, particularly among the four major domestic banks, many of which are business process related. This has led to: i) a Royal Commission [15]; ii) regulators issuing penalties exceeding A\$2 billion[1]; iii) tightening executive accountability(See footnote 1); iv) more than A\$8 billion in remediation costs and a significant investment in compliance resources[2]; and v) the resignation of three CEOs and two Chairmen of these banks.

It is not just Australian banks that struggle. Since 2008, US banks have been fined \$243bn for compliance-related events and the global cost of compliance for financial services firms is equivalent to an 8% tax[3].

---

[1] https://www.austrac.gov.au, https://www.apra.gov.au.

[2] https://home.kpmg/au/en/home.html, https://www.robertwalters.com.au.

[3] https://www.ascentregtech.com/blog/the-not-so-hidden-costs-of-compliance/.

© Springer Nature Switzerland AG 2022
C. Di Ciccio et al. (Eds.): BPM 2022, LNBIP 458, pp. 3–20, 2022.
https://doi.org/10.1007/978-3-031-16171-1_1

Academic interest in the field of business process compliance (BPC) traces its roots back to corporate scandals at organizations such as Enron, HIH, AIG, Lehmann Brothers and Société Générale along with the ensuing legislative changes (e.g., Dodd-Frank, Sarbanes-Oxley) at the turn of the millennium. The overarching challenge for BPC is to capture compliance requirements and check that business processes are operating in line with these requirements. This implies a need to evaluate processes throughout the BPC lifecycle: at design-time, run-time, and post-execution [14]. While research efforts have focused on automating BPC, and significant progress has been made, there are still several research gaps [14] and BPC is still highly manual and time consuming.

In this setting, this paper investigates the reasons why banks struggle to keep up with BPC. We focus on the Australian banking context, and then discuss how the findings are generalizable to other banking contexts.

To this end, first, we conducted a series of semi-structured interviews on BPC and its challenges. Next, following the Gioia methodology [12], we thoroughly analyzed the interview transcripts to identify factors inhibiting BPC. The interviews were conducted with participants drawn from the four major Australian banks, namely Australia and New Zealand Banking Group, Commonwealth Bank of Australia, National Australia Bank, and Westpac Group. The participants had backgrounds in *Operations*, *Risk & Compliance*, *Technology*, and *Process Excellence*.

In light of the above, this paper contributes a conceptual framework that identifies and aggregates 23 concepts capturing various challenges emerging from the interviews into seven key *themes*, which are further grouped into three *aggregate dimensions*. These dimensions are symptomatic of a fail-fix cycle that is entangling the banks. Based on this, the paper further suggests a circuit-breaker approach to address this cycle.

The remainder of this paper is structured as follows. Section 2 provides background to the study, including the regulatory context for Australia's major banks and the relevant BPC literature. Section 3 outlines our research methodology, results, and analysis. Section 4 discusses our findings while Sect. 5 presents the limitations of the study. Section 6 concludes the paper and discusses avenues for future work.

2   Background and Related Work

To address our research question, an understanding of both the BPC literature and the Australian banking context is required. Here, we provide a summary of both.

2.1   Business Process Compliance

BPC is a well researched area. A BPC solution comprises multiple elements, each with a range of techniques proposed in the literature. At its core a BPC solution must demonstrate: an ability to capture requirements [8]; an approach and a language to formalize the rules [11,13]; an approach to represent the process [4,24]; a technique to check compliance between the process and the rules [9,19] with regards to different process perspectives [17]. There are also a range of supporting features that enhance BPC's value such as: business reporting [24]; violation handling [22]; feedback and root cause analysis [25]; and change handling [21]. Most contributions focus on one or more of these elements at a specific stage of the BPC lifecycle: *design-time*; *run-time*; and *post-execution* (i.e., auditing) [14].

Framework-oriented solutions provide the backbone for BPC efforts [14] and cover a broad spectrum, from enterprise-wide, high-level risk management frameworks (e.g., COSO) to industry and function specific frameworks (e.g., Basel accords) to frameworks that aim to solve a specific piece of the BPC puzzle, e.g., a taxonomy-based framework in [26], or an evaluation framework in [20].

Managing BPC at design-time is a preventative strategy, concerned with ensuring that processes comply with relevant rules and regulations prior to execution – either during the design-process [13] or post-design but pre-execution [23]. Debate has centred on approaches and languages that are expressive enough to handle the range and complexity of compliance requirements [13] but are seen to be technically complex, and those languages that are more business-user friendly (e.g., pattern-based approaches) but potentially lack some of the expressiveness [11]. A range of techniques have been used to represent the process, such as Petri nets, UML diagrams, BPEL models, however, BPMN models are becoming the most popular in industry.

Run-time methods verify compliance during the process execution, and typically address aspects of BPC that cannot be verified and validated at design-time, e.g., "segregation of duties" and "deadlines for completion" requirements [22]. Proposed solutions fall into two broad categories: reactive, where compliance verifies progress-to-date [22]; and proactive monitoring, where progress-to-date knowledge is used to predict compliance outcomes [18]. While there are a range of techniques to capture the run-time process data, event streams are becoming the dominant approach [18,22]. However, this is challenging for processes producing sparse event streams, and computationally complex for processes producing large amounts of events in a short time.

Auditing is a post-execution strategy, traditionally both manual and based on sampling, there is now a shift to continuous auditing. Some of the auditing approaches covered in the BPC literature are based on process mining (PM) techniques [9,25,27], these techniques benefit from reviewing a population of transactions instead of a sample. Database-driven solutions have also been proposed [1,16]. As a detective control, auditing does not prevent compliance breaches, but can be useful to inform process enhancements and also assess the impact of changed requirements.

While much progress has been made in BPC research, there are outstanding challenges to apply the techniques in real world scenarios [6,14]. There is a recognition that the goal of automating BPC may be out of reach, with the focus now shifting to facilitation rather than full automation [5].

2.2 Australian Banking Context

There are 98 banks operating in Australia, controlling A\$5.2 trillion of assets[4]. The four major banks account for 74% of these assets. Over the last five years, multiple compliance breaches have been made public, highlighting a range of challenges the banks face in trying to maintain process compliance.

The findings of the Royal Commission and other regulators include: the extent of legislation banks are subject to; the difficulties banks have both understanding and interpreting them; blurred lines of accountability and bureaucratic decision-making;

---

[4] https://www.apra.gov.au.

the extent of processing and administrative errors (A\$239m repaid in mortgages alone); poor processes; the age and complexity of product systems; a reactive approach to operational risk management and inability to detect systemic issues; a reliance on manual, detective controls that do not operate end-to-end; issues not addressed in a timely manner; trade-offs between funding compliance initiatives versus other initiatives; and an inability to "join the dots" [2,3,15,28].

3   Methodology, Results, and Overview

In this section, first, we introduce the Gioia methodology [12] and discuss how we applied it to the context of this study. Then, we report the results we obtained and provide a broad overview before discussing the results in depth in the next section.

3.1   Methodology

In this study we applied the Gioia methodology [12], given its ability to bring "qualitative rigor" to the conduct and presentation of inductive and abductive research. It provides guidelines to create a conceptual data structure comprising *1st Order Concepts* directly extracted from a set of interview transcripts, then analyzed and consolidated into *2nd Order Themes*, and finally distilled into *Aggregate Dimensions*. Specifically, we executed the Gioia methodology by completing the following seven steps.

1. Develop Interview Protocol. First we developed an *interview protocol* to conduct the interviews. Development of the protocol was informed by a review of the BPC literature and enriched by an understanding of the banking context. It started with three introductory questions which sought participants views on what BPC meant, the impact of the recent regulatory issues, and which teams were impacted. The heart of the interview focused on: examples of the BPC issues the participants experienced, what they would do in hindsight, how they thought the issues could be prevented, how they would measure BPC performance, and the role they thought process mining could play to address the issues.

2. Select Interviewees. The industry participants in this study (i.e., the interviewees) were drawn from the authors' network. 54 potential interviewees were approached and 22, one-hour, semi-structured interviews were conducted. The interviewees were predominantly Senior Managers, Heads of, and General Managers within the relevant organization. All had banking experience in the last five years with at least one of the four major Australian banks. The average banking tenure (including international banking experience) was 17 years, and half of the participants had worked for more than one of the banks. Each of the banks was represented by at least five interviewees who had worked there. The interviewee profile is shown in Fig. 1.

3. Conduct and Transcribe Interviews. We sent a *Plain Language Statement* to participants prior to their interview, to provide them with context. All the interviews were conducted over Zoom calls, between June and December 2021. Initial pilot interviews were conducted with eight industry participants, to validate the line of questioning. Each recorded interview was subsequently de-identified and transcribed.

4. Code Transcripts. The interview transcripts were imported into Nvivo and a word frequency count was run on the interviewees' responses to identify key terms based on both exact and stemmed word matches. The results were mixed. Some of the most frequently used terms had ambiguous meaning. For example, the word "end" was frequently referenced regarding end-to-end process – a salient term in this study, but it was also used as a figure of speech, e.g., "at the end of the day", "the end result".

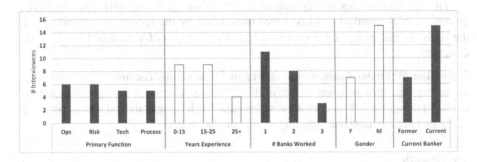

Fig. 1. Interviewee Profile

Hence, while the automated word frequency functionality provided some insight, its usefulness was limited to providing a base list of frequently used terms. This was addressed by manually coding the derived list of terms to each paragraph of the pilot interview transcripts, where they were used in the relevant context. The terms were also enhanced with participant-used synonyms. The result of this step created 152 codes, with some paragraphs associated with as many as 22 codes.

The remaining 14 transcripts were subsequently coded. During this process an additional 17 terms were identified and added as new codes. No new codes were added for the last three transcripts suggesting saturation had been reached. Each transcript was coded to between 100 and 121 codes, with an average of 111 codes. The number of references per transcript ranged between 401 and 734 with an average of 511.

5. Develop 1ˢᵗ Order Concepts. Following the initial coding, we then derived the 1ˢᵗ Order Concepts through an iterative process. The 152 codes were consolidated by relying on the available data and the expertise of the authors of this paper. The first task was to consolidate terms into synonymous concepts. For example, terms such as "audit", "QA, QC" were consolidated with terms like "controls" and "checks". The results were reviewed by two co-authors, who provided suggestions for refinement, and then another iteration would start – until no other changes were proposed. This iterative approach exposed two main concerns. First, some terms were applicable in a range of contexts, e.g., the most frequently used term was "process", but depending on the context it could refer to process design, a specific banking process such as mortgage lending, or process mining. Second, not all references to a term were indicative of a BPC challenge, e.g., some interviewees referred to a term in a favorable light. Taking these points into consideration, further iterations resulted in the terms being consolidated into 23 1ˢᵗ Order Concepts.

6. Validate Coding. To validate the coding, four transcripts were randomly chosen and coded separately to the 1st Order Concepts by two authors. Only the interviewee's comments that referred to a concept as a challenge were coded. The results were compared for consistency, calculating the *Cohen Kappa Coefficient* (CKC) for each transcript [7]. The CKC measures inter-rate reliability allowing for chance agreement. The resulting CKC, for each of the four transcripts, was between 90% and 93%, suggesting that there was a high degree of consistency.

The few mismatches mainly related to the authors' different expertise and background. A final round of validation was undertaken by the remaining two authors of the study. In this case, they reviewed selected text extracts for each of the 1st Order Concepts.

7. Develop 2nd Order Themes and Aggregate Dimensions. The final stage to generate the output *data structure* was to derive the 2nd Order Themes and Aggregate Dimensions. To do so, we followed an iterative process that leveraged the industry experience of the first author[5] alongside existing theory and literature.

3.2 Results

We describe the results in three layers: 1st Order Concepts, 2nd Order Themes, and Aggregate Dimensions. We begin with a description of the 23 1st Order Concepts, and

Table 1. Resulting *Data Structure* from our analysis. The last two columns respectively report the references per concept and the number of interviewees referencing a concept.

1st Order Concepts	→	2nd Order Themes	⇒	Aggregate Dimensions	# Ref.	Int.
Complex Model	→	Dynamic & Complex Ecosystem	⇒	Complex & Dynamic Requirements Load	72	21
Regulatory Pressure Intensifying	→				84	18
Disruptive Competition	→				19	10
Frequently Changing Direction	→				22	9
Multiple Requirement Types	→	Complex Requirements			30	20
Translating Ambiguous Requirements	→				37	14
Conflicting Objectives	→				64	19
Inflexible, Disconnected Legacy Technology	→	Disjointed & Disparate Process Foundations	⇒	Impenetrable Spaghetti Processes	71	21
Data Oasis, Information Mirage	→				66	17
Fragmented Processes	→				72	19
Inadequate & Ineffective Support	→	Hard to Follow Processes			61	16
Too Many Exceptions	→				38	17
"Band-Aids", Patches & Workarounds	→				51	19
Huge Scale	→	Resource Intensive Processes, Prone to Fail			18	12
Partial Automation Relies on People	→				72	21
Layers of Flawed Controls	→				106	21
Lack of Knowledge & Experience	→				117	20
System Monitored Not Processes	→	Decision-Making Blind Spots	⇒	Elusive Case for Sustainable Change	29	13
Impaired Line of Sight	→				92	20
Change Execution Credibility	→				82	20
Short-Sighted Investment	→				43	14
Unclear Accountability	→	Cultural Headwinds			86	20
Tick-the-Box Culture	→				78	20

---

[5] The first author is a former Senior Executive of two of the four banks studied, with more than 15 years of experience.

then discuss how we aggregated them into the 2nd Order Themes and the Aggregate Dimensions. The data structure summarizing our results is reported in Table 1.

1) Complex Model. The banking business model is complex. There are a large number of products and services offered through multiple channels to a wide-range of customer segments across multiple jurisdictions, which are organized around multiple business units as part of an ecosystem dependent on many and varied 3rd party stakeholders. It is not just the dimensions of the model but the interconnected web that they form, e.g., products and processes that cut across organizations, business units and segments, and that are largely invisible and intangible. This complexity translates into a significant number of requirements that must be captured in process design.

2) Regulatory Pressure Intensifying. Interviewees focused on the changing nature of regulatory relationships, larger fines, more requirements and increasing scrutiny, or as Interviewee-1 put it: *"[...] it's just a wave after wave of regulation"*. Interviewees also referenced: the cost of compliance impacting competition; the regulators' product knowledge limitations; the fact that the burden of compliance is being felt directly by customers; the expanding role banks are expected to play helping police financial crime; and the perception that the regulatory relationship with the banks is adversarial, whereas a more collaborative approach is required to address many of the industry's issues.

3) Disruptive Competition. Changing industry dynamics are also creating more requirements. New players are not burdened by inflexible technology, are far more agile – introducing new products and features at a faster rate – and in some cases are more lightly regulated. This is seen to present a cost advantage to non-bank participants but an increase in risk to the overall system.

4) Frequently Changing Leadership and Direction. Staffing, structure and strategy changes, particularly at senior leadership levels, create work. It is not restricted to major changes but also more routine business decisions such as individual leaders changing roles, reducing project budgets or changing their risk appetite. These lead to a change in objectives, projects being re-scoped in-flight and resources re-distributed.

5) Multiple Requirement Types. In terms of the types of requirements that must be fulfilled, the initial response, for almost all participants, was to focus on regulatory obligations and requirements. However, follow-up questions revealed that there are many other types of requirements: industry codes of practice and standards; business policy; contractual obligations; and, of course, customer requirements. Variation also affects the necessity of a requirement (e.g., mandatory or "nice to have"), as well as the consequences of failure (e.g., a significant fine or an adverse performance indicator).

6) Translating Ambiguous requirements. The way requirements are communicated is not always clear. They can be contradictory, duplicated, written in ambiguous language subject to both interpretation and translation, e.g., "We need to do the right thing". In some cases the requirements are not known or not communicated.

7) Conflicting Objectives. Interviewees referenced the focus on sales, responsiveness, service level agreement targets (SLAs), cost efficiencies, and meeting the needs of investors first as higher priorities than quality or compliance. Even the threat of larger fines is perceived to be insufficient to change the mindset. There is also conflict between different teams in a bank highlighted by Interviewee-2: *"[...] we're supposed to be innovative [...] but the brakes are put on by the compliance guys"*.

8) Inflexible, Disconnected, Legacy Technology. The technology environment comprises a multitude of in-house built systems and others sourced from multiple vendors. They are heavily customized, do not integrate easily, and are hard, slow and expensive to change, or as Interviewee-18 put it: *"We are sort of bound by the legacy system. To make a simple change in the system, it is quite difficult."* Whereas Interviewee-5 stated: *"[As a roadblock] the one that springs to mind straightaway is integration. Our technology architecture is far from simple. There's bits and pieces logged [sic] all over the place."* Interviewees also referenced individual processes dependent on 30 applications and a technology landscape with hundreds of disconnected systems.

9) Data Oasis, Information Mirage. While there is no shortage of data, classifying it and accessing it, particularly at the right level of granularity, is not easy and there are also integrity issues such as: duplication, blind-spots, missing data, or data getting lost during migrations. Without unique identifiers and standards such as naming conventions, data does not flow easily between technology assets and processes, and stitching it together is expensive and time consuming.

10) Fragmented Processes. Processes are seen as something that occur within a function, not end-to-end. As the organization changes, the process boundaries also change. Over many years this has led to significant fragmentation with bits of processes dispersed across many teams and limited understanding of how the component parts fit together. Interviewee-8 stated: *"[...] we're constantly breaking up our processes to fit with a design that's based around where people work, not what they do"*.

11) Inadequate & Ineffective Support. *"They've got really complicated checklists that they just don't use."* (Interviewee-6) is one reference to the inadequacy of the tools and documentation process participants are working with. Others include: process models developed by people with limited process modeling skills, documentation that does not exist, is not maintained or only covers the "happy path". In other cases, process maps are documented for the regulators, not for the teams operating the processes, or as an overlay, not an integral part of the process. Interviewee-11 commented: *"I know banks spend a heap of time and heap of budget on documenting processes. But if you ask the average person on the ground, they would say they're not documented."*

12) Too Many Exceptions. Referring to a lending process, Interviewee-14 commented: *"You thought you had a 70% STP [straight-through-processing rate]. The reality is you had 3% because the other 97% were taking one of the 56,000 variable pathways"*. The exceptions are typically driven by customizing siloed applications and insufficient project funding or a need to meet a customer request quickly. Exceptions are also introduced based on local considerations, e.g., the degree of latency or staff trying to navigate an easier path through the process.

13) "Band-Aids", Patches & Workarounds. Years of under investment and constant tweaks, tinkering and point solutions have left processes strewn with workarounds, patches, bottlenecks and hand-offs. With a *"don't fix'til it's broken"* mindset

(Interviewee-10), the ensuing urgency leads to more patches and workarounds. It also contributes to high rework rates and errors as the processes do not flow smoothly. Inadequate workflow was raised by multiple interviewees. Some refer to a workflow based on email and collaboration tools, others comment that only part of the end-to-end process has been workflow-enabled, negating the benefit.

14) Huge Scale. While the scale varies by process, the high volume of time-critical transactions is frequently referenced, e.g. payments. Seasonality effects are also prevalent. It is not just high volumes; interviewees referenced the number of staff – between 30,000–50,000 per bank, instances of putting hundreds of risk controls, and concurrent onboarding of 90+ new employees into a single team.

15) Partial Automation Relies on People. High-volume processes are only partially automated, resulting in a significant number of manual, repetitive tasks. It is not just the higher likelihood of errors, but the ability to absorb the degree of change and the lack of audit trails/visibility that prove challenging. Interviewees see automation as a cure-all.

16) Layers of Flawed Controls. The second most referenced concept, adding more controls is seen as the response to any problem. Many controls are after-the-fact and manual, many are flawed (e.g., 4-eye checks), many rely on sampling. Some are so complicated that they are not applied and the layering of controls mean that many are never activated. The lack of preventative controls is attributed to the fact they would slow the process down by multiple interviewees.

17) Lack of knowledge & experience. The most frequently referenced 1st Order Concept is driven by: a loss of knowledge and experience as long-tenured staff leave; the narrowing of focus to learn discrete tasks instead of the end-to-end process; the difficulty in attracting talent to work with legacy technology; the time it takes to train new people (up to twelve months); and the workload pressure the teams face.

18) Systems Monitored, Not Processes. Business activity monitoring and systems monitoring are referenced by multiple interviewees, but while they may trigger alarms in terms of queue depth and system performance they provide no indication of process performance. Interviewee-3 commented: *"So most of our [operational technology] monitoring has a focus on system health and availability [...] in terms of monitoring actual processes, we don't really have that in place."*

19) Impaired Line-of-Sight. Metrics and reporting do not link business outcomes to process performance and systems events. Different teams look at different metrics, with different objectives, hence decision-making is challenging. Some interviewees believe the data is there, but it has never been considered important enough to extract. Some assert the data exists but does not translate into decision-making information. It is not just about availability, but also about actionability for real-time decision-making.

20) Change Execution Credibility. With processes so dispersed and so many stakeholders, getting buy-in, managing the various self-interest groups and capturing requirements up front is challenging. Communicating the change and the implications are also seen by interviewees as gaps. Interviewees referenced projects running over budget and then being re-scoped, typically leading to more workarounds. More worryingly were "improvement" projects that made things more complicated, unwittingly removed key controls, did not deliver or did not actually fix the problem. As Interviewee-19 noted: *"I've been here 11 years and I have not known one [of 20 projects] fix the problem statement that we need to fix."* Given the change process can be slow, unofficial, shadow change processes also play a role – which leads to more tinkering.

21) Short-Sighted Investment. The reason for not progressing the business case is typically referenced as: the investment is too high, the time to make the changes too long, the benefits and potential value are unclear and too far in the future when results are needed now. In other words, the sustainable, strategic business case does not stack up relative to the alternative of more FTEs (Full Time Equivalent resources), patches and remediation. Interviewee-11 added an additional insight: *"Senior leaders [...] want to work on the strategic stuff. I'm not sure processes are seen as strategic enough."*

22) Unclear Accountability. Accountability is confused, particularly with regards to the three lines of defense model. One incident referenced involved an account owner, an ATM network owner, a branch owner, a product owner, and a customer owner. There was no reference to a process owner. Progress has been made, while the BEAR[6] legislation does not attribute process ownership directly, it at least makes it implicit, and there has been investment in bolstering risk and compliance support resources. However, interviewees believe that this has led to accountability being removed one step further from the source of risk. The propensity to engage more consultants and lawyers has had the same effect.

23) Tick-the-Box culture. When asked who is responsible for compliance, "it's everyone's job" was a common response, with the heightened scrutiny creating a sense of nervousness. However, the message becomes confused and appears to lose momentum as it filters down through the organizational hierarchy. Throughout the transcripts there are references to BPC being perceived as a toll-gate, a "tick-the-box" exercise, people mechanically following a process whether it is right or wrong to avoid blame – even though consequence management is rare – people not challenging the status quo or afraid to speak up and BPC tasks executed with a sense of complacency. It is not seen as strategic but it is important to be seen to act.

Synthesizing these 23 *1st Order Concepts* together led to the identification of seven *2nd Order Themes*. The complex business model, intensifying regulatory pressure, disruptive competitive landscape and the impact of leadership and directional changes combine to highlight a *Complex & Dynamic Ecosystem* generating a significant number of *Complex Requirements* – there are many different types, they are frequently ambiguous, and they must cope with conflicting objectives and priorities. This requirement load is imposed on *Disjointed & Disparate Process Foundations*, where the underlying legacy technology is poorly integrated and inflexible, the data is plentiful, but hard to access and limited in its ability to convey information, and processes are repeatedly re-aligned to follow organizational structure changes. Additionally, the lack of support materials and the extent of exceptions and repeated patching lead to *Hard to Follow*

---

[6] The Banking Executive Accountability Regime establishes accountability obligations for senior bank executives and directors.

Processes. Despite such challenging foundations, the volume going through the processes is significant, yet they are only partially automated and not error-proofed, hence they are *Resource Intensive Processes, Prone to Fail*, exacerbated by the fact that there is a lack of knowledge and experience. Not monitoring end-to-end processes thwarts the ability to align process performance and business objectives, and with a poor track-record executing change these *Decision-Making Blind Spots* hinder investment. There are also *Cultural Headwinds*, where ownership and accountability for addressing the issues is unclear and BPC is seen as a "tick-the-box" exercise, not a strategic endeavor.

These seven 2nd *Order Themes* were further consolidated into three *Aggregate Dimensions*. A complex and dynamic ecosystem coupled with a high degree of requirements complexity leads to a *Complex & Dynamic Requirements Load*. This demand on the organization is fulfilled by *Impenetrable Spaghetti Processes*, where resource intensive processes are hard to follow, prone to fail and built on disjointed and disparate process foundations. Decision-making blind spots across end-to-end processes and cultural headwinds make for an *Elusive Case for Sustainable Change*. Together, these dimensions provide insight into why the major banks find BPC challenging.

3.3   Overview

The purpose of this study is to understand why banks find BPC challenging. It should be noted that bank processes are not homogeneous and interviewees also provided examples where some of the 1st *Order Concepts* were being, or had been addressed. However, for those processes experiencing BPC challenges, the results above represent the common themes.

From this point on, we refer to individual 1st *Order Concepts* with a "C" and their ID number as listed above. Overall, 48% of references were associated with the *Impenetrable Process Spaghetti* aggregate dimension, 29% with *Elusive Case for Sustainable Change*, and 23% with *Complex & Dynamic Requirements*. The number of references per 1st *Order Concept* ranged from 18 to 117 (see also Table 1), with the top five concepts accounting for 34% of the references (C17, C18, C19, C22, C2 respectively) and the top eleven accounting for 66%. 13 1st *Order Concepts* were referenced by more than 18 of the interviewees, and only two concepts were referenced by less than half of the interviewees (C4 and C3 respectively).

In terms of coding differences, with a limited number of exceptions, interviewees' responses were relatively homogeneous. There were no material differences (within ±5% of the average) by gender and current role. Those with more than 25 years of tenure focused less on *Impenetrable Spaghetti Processes* while those who had worked in three of the banks placed more emphasis on *Complex & Dynamic Requirements Load* (+8%) and less on *Elusive Case for Sustainable Change* (-9%). By function, Operations and Process Excellence interviewees focused more on *Impenetrable Spaghetti Processes* (+8% and +6% respectively). The Risk & Compliance interviewees emphasized *Complex & Dynamic Requirements Load* (+10%) and Technology interviewees favored *Elusive Case for Sustainable Change* (+7%). This was offset by Risk & Compliance and Technology interviewees placing less emphasis on *Impenetrable Spaghetti Processes* (−6% equally) and the Process Excellence interviewees focusing less on *Complex & Dynamic Requirements Load* (−6%).

The implications of these results are discussed in the next section.

4   Discussion

Some of the challenges associated with the 1st Order Concepts may sound familiar. The BPC literature refers to: the difficulty in extracting and translating ambiguous requirements (C5, C6); the importance of process ownership and clear roles and responsibilities (C22); aligned metrics and reporting (C19); managing scale (C14); incorporating real-time monitoring (C18); the importance of managing change well (C20); the importance of proactive, preventative controls and compliance-driven design (C16); and the relative merits of annotating (overlaying) versus integrating controls in process models (C11). The business process management (BPM) community is more than familiar with the impact of variation (C12); the importance of good documentation (C11); adequate training (C11); fit for purpose automation (C15); legacy integration (C8); and data-connectivity issues (C9).

The BPC literature proposes many solutions to (partially) address these issues, and it is tempting to suggest appointing a team of process professionals to at least fix the process-related issues. However, as Interviewee-8 pointed out: *"Every three to five years the banks get rid of all their process improvement people."* Why is this the case?

Interviewees referenced the invisible, intangible nature of processes in banking, making it harder to see when and where a process has failed. Others referenced the fact that the banks do not control the end-to-end process (e.g., in payments and broker originated mortgages), hence, managing input quality with ecosystem partners is harder. However, there are examples of solutions cited in the literature referencing common bank processes – the same processes discussed by the interviewees: account opening, lending, payments, and customer onboarding. Another interviewee referenced the fact that there are risk people and operations people but no risk operations people. Again, the BPC literature assumes that these are people with separate skill sets working collaboratively, as has been the case in the major banks, so there should be no impediment.

The answer may lie in four of the most referenced terms: "end-to-end", "complexity", "perspective", and "understand" with 154, 141, 205 and 270 references respectively. The terms appear in multiple contexts, which have been assigned to the most relevant *1st Order Concept*, but the overarching theme is that, from the interviewees perspective, end-to-end processes are so complex that people do not understand them. Interviewee-1 summed it up by saying: "So I think this spaghetti, this complexity that underpins very aged infrastructure with Band-Aids plastered across it. The lack of knowledge. Who understands that [...] when it falls over? And people apply another fix and another fix." Interviewee-11 elaborates on this: "I think complexity comes from [...] the fact that no one knows what's going on", while Interviewee-8 states: "Most people [...] do not understand the fundamentals of what a process is" and also refers to a short-sighted view of process.

This is supported by interviewee's references to the inadequacy of the tools and techniques used to help people understand their processes in the identification, discovery, and analysis phases of the BPM lifecycle. Examples include: process maps that do not reflect reality, a lack of modeling skills, a reliance on subject matter experts who do not understand their processes nor their requirements, time constraints that mean the exceptions are not mapped, and the lack of performance data to inform the analysis of the current state. Our observation is that interviewees find it hard to see and analyze the system holistically, and people resort to solving in their own silos.

Moreover, when tasked with improvement, interviewees referenced three pathways: i) the complexity is underestimated in the project and the cost and timelines blow out, so the project is shut down; ii) the complexity is acknowledged in the project, cost is incurred but results are not delivered in a timely manner, so the project is shut down; and iii) an improvement to part of a process variant that is so immaterial no one notices.

Investment in a sustainable solution appears to be elusive. This can be attributed to: limited progress-to-date, "siloed thinking", an inability to "see" the intrinsic, system-wide costs of the spaghetti processes, the difficulty in "selling" this type of business case, the level of investment required to fix it and a nebulous benefit case. Hence, the fail-fix approach persists, considering also the excessive requirements' load, the complexity increases. This fail-fix cycle emerged from the analysis of the interviewee transcripts. Our conclusion is that, while there may be solutions to the 1st Order Concepts individually, an effective solution requires a more holistic perspective, and hence a shift to seeing those first order challenges through the lens of the 2nd Order Themes. We elaborate this perspective in Fig. 2.

Our analysis of insights from the interviewees points to three fundamental challenges to break the cycle: i) end-to-end visibility; ii) effective collaboration across the ecosystem to simplify requirements; and iii) developing a politically viable, sustainable business case. In the following, we comment on each of these circuit breakers.

End-to-End Visibility. The first circuit breaker is to focus on end-to-end visibility. As Interviewee-8 put it: *"If you can't see into your process, [...] you're running blind."* Banks must be able to see the complexity to chart a course to untangle the spaghetti. However, this is not a traditional process discovery and analysis approach. The limitations discussed above preclude the traditional approach in favor of *automated process*

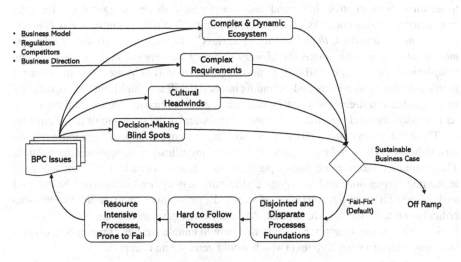

Fig. 2. BPC fail-fix cycle

discovery [10], a core process mining capability. Seeing all the exceptions, all the pathways, all the performance data associated with those pathways is essential to improve transparency, enable visibility and enable analysts to view the system holistically. Most interviewees were enthusiastic about the approach, and in some cases pilot process mining projects were already underway.

The roadblock, as many interviewees pointed out, is accessing the event data across end-to-end processes and addressing the data blind-spots when partially automated processes reverted to manual tasks. Extract, transform and load (ETL) techniques and tools have certainly advanced in recent years, but stitching together data without a common identifier, with blended coarse- and fine-grain events, will require further work.

Assuming reliable end-to-end event logs can be generated, a criticism of the automated process discovery approach is that it tends to produce spaghetti models – 56,000 paths through a lending process is a case in point (Interviewee-14). While it is essential to be able to see the full complexity, it is critical that techniques and tools also simplify abstraction. Interviewees refer frequently to *layers* and *perspectives*. The Risk & Compliance function want to see the process through the controls lens or the regulatory lens, operational teams want to see the resource impacting flows and queues/bottlenecks. Our observation is that automated process discovery tools should provide a single model, capable of capturing the different user perspectives.

Interviewees reflected that leveraging event logs presents a range of other process mining opportunities, particularly, online conformance checking and variant analysis would help address C16 and C12, respectively, while automated process discovery also enables the other circuit breakers.

Effective Collaboration Across the Ecosystem to Simplify Requirements. The second circuit breaker is to simplify requirements. At present, requirements are treated in silos, (e.g., regulatory requirements, codes of practice requirements, business policy and customer requirements). However, there is a significant amount of overlap. A small number of control patterns [11] can be implemented to cover the majority of requirements. Because there are so many of them that are added incrementally, it is difficult to see through the control clutter and understand which ones are triggered in which circumstances and sequence. As Interviewee-15 put it: *"You'll have multiple risk controls that are never activated, therefore they're useless"* and they go on to say *"I think the more controls you put in place, the higher the risk of any process, because you can add complexity"*. End-to-end visibility as described above will help users see the requirements complexity as well as understand its impact. It will also highlight the opportunity to standardize requirements registers, and risk & control libraries. Hence, making it easier to consolidate and prioritize rules and controls across the portfolio of requirements.

There is also an opportunity to work with ecosystem participants including regulators, industry bodies, and competitors to support simplification through standardization. Throughout the interviewee transcripts there are references to a lack of standards such as naming conventions and acronyms and terminology applied inconsistently. As well as developing a more rigorous suite of standards, taxonomies and ontologies, it is also critical to ensure that they are applied. An interviewee gave an example where member banks could choose to apply a code (which would enable straight-through-processing) but many chose to enter free text (which would generate an exception).

Developing a Politically Viable, Sustainable Business Case. End-to-end process visibility also enables the third circuit breaker – developing a politically viable business case for sustainable change. The fail-fix cycle is perpetuated not because of a lack of funding and resources, but because the business case to address the issues sustainably does not *appear* to be politically viable. Interviewees refer to the level of investment required, the fact that senior leaders do not perceive BPC to be strategic, the lack of process ownership (sponsorship), that product and process costs are not known and the value of doing this well and sustainably are unclear. Leveraging the enhanced end-to-end visibility, the task is to demonstrate the value of a strategic approach to BPC: the impact of exceptions on the level of operating cost and risk capital held, the cost of re-hiring and training people on broken processes, the opportunity cost of not being able to absorb more change, the "tech debt" associated with hollowing out legacy systems and keeping them on life support, the impact on product pricing of actually knowing your costs. It is hard to answer these questions accurately in the major banks today.

The second part to this circuit breaker is to package up programs of work that can deliver value in the financial periods the banks are beholden to. We asked interviewees where would they start if they were in charge of addressing the issues. The majority focused on piggy-backing off existing programs of work. They would select a process that was material to the bank's results, a significant pain-point for executives, one that is fully funded and with a senior executive sponsor already in place. They were adamant it should not be a program in its own right. The objective should be to enable an existing program to deliver better outcomes faster by addressing the issues discussed.

5 Limitations of the Study

There are several potential limitations of this study: we only interviewed 22 current and former staff members of Australia's four major banks. These banks collectively employ over 150,000 FTEs. However, the interviewees represent key categories of process compliance stakeholders identified in the literature, namely Operations, Risk & Compliance, Technology and Process Excellence. It is also important to note that saturation was reached after 19 interviews.

In selecting our sample we wanted to ensure that interviewees were close enough to the point of execution to produce detailed anecdotes, but not so close that they could not see the broader context. To this end, the study focused on senior management. The executive layers, we determined, were too far from the detail and the process operators' focus was too narrow. Extending the study to cover other perspectives within the hierarchy is a potential future work stream.

This case study focused on four Australian banks. There is a risk that the findings cannot be inferred for other banks either within Australia or in other jurisdictions or other financial services participants. However, the Royal Commission found similar issues across the broader Australian financial services industry and other countries have also experienced similar situations. While significant effort has been undertaken to validate the results as described in the approach, there is still a degree of subjectivity in interpreting the findings, where the experience of the first author was relied on.

6 Conclusion

Through a series of semi-structured industry interviews, our study has identified 23 1st Order Concepts, linked to seven 2nd Order Themes and three Aggregate Dimensions that help explain why major Australian banks find business process compliance so challenging. Challenges associated with some of the 1st Order Concepts are known and well researched. The balance tend to be seen as organizational issues beyond the scope of this type of study. The most frequently mentioned concerns refer to i) a lack of knowledge and understanding of people involved in trying to establish and maintain business process compliance, followed by ii) a lack of preventative controls, and iii) a lack of visibility across end-to-end processes. However, our findings suggest that treating each of these concerns as a stand-alone issue will not address the underlying problem – the major banks have already tried this approach.

What makes our study different is the focus on addressing the overarching issue of the *intrinsic complexity*. The current approach of the banks, particularly how they approach process discovery and analysis, locks them into a negative fail-fix cycle of increasing complexity, which demands a circuit-breaker. We propose three areas to focus on to break the cycle: i) enhancing end-to-end visibility through automated process discovery; ii) simplifying requirements across the ecosystem; iii) and developing a politically viable, sustainable business case.

It is important to note that the results of our study do not imply that every process, in each of the banks studied, exhibits all of these concerns. Counterexamples were also provided by interviewees, and each 1st Order Concept should be seen as one end of a spectrum. Future work will focus on developing a process profiling approach to help banks determine the extent of the business process compliance challenge by process. Further work is also required to determine the role of specific process mining techniques such as conformance checking and variant analysis and how they can address the challenges identified in this study.

Acknowledgments. We thank the interviewees who generously gave their time and shared their thoughts to help produce this paper.

References

1. Agrawal, R., Johnson, C., Kiernan, J., Leymann, F.: Taming compliance with Sarbanes-Oxley internal controls using database technology. In: ICDE, pp. 92–92. IEEE (2006)
2. APRA. Prudential inquiry into the Commonwealth Bank of Australia (CBA) final report, April 2018. https://www.apra.gov.au/sites/default/files/CBA-Prudential-Inquiry_Final-Report_30042018.pdf
3. ASIC. Rep 594 review of selected financial services groups' compliance with the breach reporting obligation (2018). https://download.asic.gov.au/media/4879889/rep594-published-25-september-2018.pdf
4. Awad, A., Decker, G., Weske, M.: Efficient compliance checking using BPMN-Q and temporal logic. In: Dumas, M., Reichert, M., Shan, M.-C. (eds.) BPM 2008. LNCS, vol. 5240, pp. 326–341. Springer, Heidelberg (2008). https://doi.org/10.1007/978-3-540-85758-7_24

5. Barnawi, A., Awad, A., Elgammal, A., Elshawi, R., Almalaise, A., Sakr, S.: An antipattern-based runtime business process compliance monitoring framework. Framework **7**(2), 551–572 (2016)
6. Becker, J., Delfmann, P., Eggert, M., Schwittay, S.: Generalizability and applicability of model-based business process compliance-checking approaches - a state-of-the-art analysis and research roadmap. Bus. Res. **5**(2), 221–247 (2012)
7. Cohen, J.: Weighted kappa: nominal scale agreement provision for scaled disagreement or partial credit. Psychol. Bull. **70**(4), 213 (1968)
8. de Araujo, D.A., Rigo, S.J., Muller, C., Chishman, R.: Automatic information extraction from texts with inference and linguistic knowledge acquisition rules. In: WI-IAT, pp. 151–154. IEEE (2013)
9. Doganata, Y., Curbera, F.: Effect of using automated auditing tools on detecting compliance failures in unmanaged processes. In: Dayal, U., Eder, J., Koehler, J., Reijers, H.A. (eds.) BPM 2009. LNCS, vol. 5701, pp. 310–326. Springer, Heidelberg (2009). https://doi.org/10.1007/978-3-642-03848-8_21
10. Dumas, M., Rosa, M.L., Mendling, J., Reijers, H.A.: Fundamentals of business process management, 2nd edn. Springer, Heidelberg (2018). https://doi.org/10.1007/978-3-662-56509-4
11. Elgammal, A., Turetken, O., van den Heuvel, W.-J., Papazoglou, M.: On the formal specification of regulatory compliance: a comparative analysis. In: Maximilien, E.M., Rossi, G., Yuan, S.-T., Ludwig, H., Fantinato, M. (eds.) ICSOC 2010. LNCS, vol. 6568, pp. 27–38. Springer, Heidelberg (2011). https://doi.org/10.1007/978-3-642-19394-1_4
12. Gioia, D.A., Corley, K.G., Hamilton, A.L.: Seeking qualitative rigor in inductive research: notes on the Gioia methodology. Organ. Res. Methods **16**(1), 15–31 (2013)
13. Governatori, G., Sadiq, S.: The journey to business process compliance, pp. 426–454. IGI global (2009)
14. Hashmi, M., Governatori, G., Lam, H.-P., Wynn, M.T.: Are we done with business process compliance: state of the art and challenges ahead. Knowl. Inf. Syst. **57**(1), 79–133 (2018). https://doi.org/10.1007/s10115-017-1142-1
15. Hayne, K.: Royal commission into misconduct in the banking, superannuation and financial services industry. Royal Commission into misconduct in the banking, superannuation and financial services industry (2019)
16. Johnson, C.M., Grandison, T.W.A.: Compliance with data protection laws using hippocratic database active enforcement and auditing. IBM Syst. J. **46**(2), 255–264 (2007)
17. Knuplesch, D., Reichert, M., Kumar, A.: Visually monitoring multiple perspectives of business process compliance. In: Motahari-Nezhad, H.R., Recker, J., Weidlich, M. (eds.) BPM 2015. LNCS, vol. 9253, pp. 263–279. Springer, Cham (2015). https://doi.org/10.1007/978-3-319-23063-4_19
18. Leitner, P., Michlmayr, A., Rosenberg, F., Dustdar, S.: Monitoring, prediction and prevention of SLA violations in composite services. In: ICWS, pp. 369–376. IEEE (2010)
19. Liu, Y., Muller, S., Xu, K.: A static compliance-checking framework for business process models. IBM Syst. J. **46**(2), 335–361 (2007)
20. Ly, L.T., Maggi, F.M., Montali, M., Rinderle-Ma, S., van der Aalst, W.M.P.: A framework for the systematic comparison and evaluation of compliance monitoring approaches. In: EDOC, pp. 7–16. IEEE (2013)
21. Ly, L.T., Rinderle-Ma, S., Göser, K., Dadam, P.: On enabling integrated process compliance with semantic constraints in process management systems. Inf. Syst. Front. **14**(2), 195–219 (2012)
22. Maggi, F.M., Montali, M., Westergaard, M., van der Aalst, W.M.P.: Monitoring business constraints with linear temporal logic: an approach based on colored automata. In: Rinderle-Ma, S., Toumani, F., Wolf, K. (eds.) BPM 2011. LNCS, vol. 6896, pp. 132–147. Springer, Heidelberg (2011). https://doi.org/10.1007/978-3-642-23059-2_13

23. Milosevic, Z., Sadiq, S., Orlowska, M.: Towards a methodology for deriving contract-compliant business processes. In: Dustdar, S., Fiadeiro, J.L., Sheth, A.P. (eds.) BPM 2006. LNCS, vol. 4102, pp. 395–400. Springer, Heidelberg (2006). https://doi.org/10.1007/11841760_29

24. Montali, M., Maggi, F.M., Chesani, F., Mello, P., van der Aalst, W.M.P.: Monitoring business constraints with the event calculus. ACM TIST 5(1), 1–30 (2014)

25. Ramezani, E., Fahland, D., van der Aalst, W.M.P.: Where did i misbehave? Diagnostic information in compliance checking. In: Barros, A., Gal, A., Kindler, E. (eds.) BPM 2012. LNCS, vol. 7481, pp. 262–278. Springer, Heidelberg (2012). https://doi.org/10.1007/978-3-642-32885-5_21

26. Rosemann, M., zur Muehlen, M.: Integrating risks in business process models. In: ACIS, vol. 50 (2005)

27. van der Aalst, W.M.P., van Hee, K.M., van der Werf, J.M., Verdonk, M.: Auditing 2.0: using process mining to support tomorrow's auditor. Computer 43(3), 90–93 (2010)

28. Westpac. Board governance of AML/CTF obligations at Westpac: The advisory panel review, June 2020. https://www.westpac.com.au/content/dam/public/wbc/documents/pdf/aw/media/westpac-releases-findings-into-austrac-statement-of-claim-issues-media-release.pdf

On the Use of the Conformance and Compliance Keywords During Verification of Business Processes

Heerko Groefsema[1] , Nick R. T. P. van Beest[2(✉)] , and Guido Governatori[3]

[1] University of Groningen, Groningen, The Netherlands
h.groefsema@rug.nl
[2] Data61, Brisbane, Australia
nick.vanbeest@data61.csiro.au
[3] Brisbane, Australia

Abstract. A wealth of techniques have been developed to help organizations understand their processes, verify correctness against requirements and diagnose potential problems. In general, these verification techniques allow us to check whether a business process *conforms* or *complies* with some specification, and each of them is specifically designed to solve a particular business problem at a stage of the BPM lifecycle. However, the terms conformance and compliance are often used as synonyms and their distinct differences in verification goals is blurring. As a result, the terminology used to describe the techniques or the corresponding verification activity does not always match with the precise meaning of the terms as they are defined in the area of verification. Consequently, confusion of these terms may hamper the application of the different techniques and the correct positioning of research. In this position paper, we aim to provide comprehensive definitions and a unified terminology throughout the BPM lifecycle. Moreover, we explore the consequences when these terms are used incorrectly. In doing so, we aim to improve adoption from research to practical applications by clarifying the relation between techniques and the intended verification goals.

Keywords: Conformance · Compliance · Verification · Review

1 Introduction

Business process management (BPM) has adapted from supporting local rigid and repetitive units of work in factory-based processes to loosely-coupled case based processes in a wide range of different, and often regulated, business contexts. This evolution set in motion an increasing need to assess whether these business processes, supported by business process management tools, are free of error, performed as desired, and follow regulations [14]. To address these distinct—but related—issues, many techniques have been developed over the past decades to help organisations understand their processes, verify correctness and diagnose potential problems [14]. Each of these techniques is very specifically designed and tailored to solving a particular business problem or question, and may be applied at different stages of the BPM lifecycle.

In general, these techniques for verification allow us to check whether a business process *conforms* or *complies* with some specification, and often refer to the popular

© Springer Nature Switzerland AG 2022
C. Di Ciccio et al. (Eds.): BPM 2022, LNBIP 458, pp. 21–37, 2022.
https://doi.org/10.1007/978-3-031-16171-1_2

business process mining technique *conformance checking* and the verification of *regulatory* compliance in BPM. While there are surface similarities among the verification problems and the activities specific to them, the terms have distinct meaning in the area of verification and their use depends on whether only specifications or a specification and implementation is involved in verification [15]. In everyday language, however, the terms conformance and compliance are often used as synonyms, and their distinct differences in verification goals is blurring. As a consequence, the terminology used to describe the techniques or the corresponding verification activity does not always match with the precise meaning of the terms as defined in the area of verification.

Due to the duality of the use of the conformance and compliance terms, several issues have emerged. In science, the confusion of these terms has lead to (i) the wrong motivation being given to justify the work, (ii) a wrong example being used to explain the work, (iii) discussions of related work including irrelevant and excluding relevant work, or (iv) evaluations comparing tools related to different perspectives. Moreover, in practical settings the confusion of these terms may lead to (v) the wrong approach being chosen and answering a question from a different perspective, (vi) the wrong artifact being used for an approach, or (vii) the approach being performed at the wrong stage of the BPM lifecycle. As a result, this inadvertently emerged mismatch between techniques and terminology could harm transfer from research to practical applications, possibly stagnating adoption of relevant approaches and new advances in the field.

In this position paper, we aim to provide comprehensive definitions of the two notions, describe the activities related to them, and the BPM artifacts they apply to.

Method and Structure

To do so, we first define the key artifacts in the BPM lifecycle and introduce the concept of verification in that context and the verification corresponding relations in Sect. 2. Subsequently, we explore the existing goals of verification and the related verification techniques for each goal in Sect. 3, discussing the intent and constraints of each verification goal. Note that, as we define each of the above elements, many definitions refer to the *ISO/IEC/IEEE Systems and software engineering – Vocabulary* standard [17]. Since the vocabulary lists multiple alternative meanings of each term depending on its application domain, throughout this paper we either directly use the variant that relates most to the domains of verification and business process management, or a combination of relevant variants. We do so, because these variants offer the best foundations required for the discussion around the verification of conformance and compliance within the BPM lifecycle. Next, Sect. 4 uses the provided definitions of artifacts and relations to connect verification relations to verification goals and provide a structured overview, highlighting potential areas that may cause confusion and propose a solution. Section 5 discusses the relevance of such a solution by providing examples of terminology and verification goal mismatches and discussing potential consequences. Finally, the findings are summarised in Sect. 6.

2   Verification and the Business Process Management Lifecycle

Validation and verification are well-known evaluation procedures used to investigate whether a software or hardware product fulfills its intended purpose [17]. *Validation* investigates whether the product fulfills the needs of the user, that is, it tries to answer if the correct product is being made. *Verification*, on the other hand, investigates if the product matches with its specifications, or whether the product is being made correctly.

When applying formal methods of mathematics to verification, the procedure is called *formal verification*. Formal verification entails proving or disproving the correctness of a model with respect to a *specification* using formal methods of mathematics. In this case, the model is a representation of the actual system (e.g., based on a specification), just like a business process model is a representation and specification of the actual business process that is being performed. Note, however, that given different verification approaches the model is not necessarily always represented by a business process model. In fact—as we will observe later—sometimes the business process model represents the specification of the verification approach instead. A specification is defined as follows:

Definition 1 (Specification). *A collection of statements that specify in a complete, precise, and verifiable manner, the requirements, design, behavior, or other characteristics of a system or component, and—often—the procedures for determining whether these provisions have been satisfied [17].*

The procedure of verification is an important aspect of the BPM lifecycle [6]. An overview is given in Fig. 1, where we map the business process artifacts of the lifecycle—represented by the circles—with the verification techniques, represented by the arrows connecting different artifacts. For each verification technique, the artifact used as the specification is connected to the artifact used to represent the model using an arrow. For example, the design properties (specification) are verified against the business process model (model) when checking process correctness. For completeness, two dashed arrows representing the validation relations have also been included. These relations are outside of the scope of this

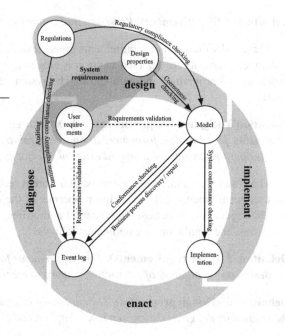

Fig. 1. Verification techniques applied during the BPM lifecycle.

paper, which has a focus on verification within the BPM lifecycle alone.

2.1 Business Process Management Artifacts Used for Verification

The BPM lifecycle uses and produces a number of artifacts that can be applied as the model or specification during a number of useful verification techniques. We define and discuss the relevant artifacts depicted in Fig. 1 as circles.

Before defining the relevant artifacts, however, we must first define the business process itself. Informally, a business processes is a collaborations between actors that achieve a specific value-added goal. Within a business process, actors perform activities based on available data and using available resources. When referring to a business process, we refer to the real life process—which may or may not be supported by software systems or be described by a model. More formally:

Definition 2 (Business process). *A partially ordered set of activities, performed by actors using available resources and data, that achieve some desired objective of an organization [17, 19, 20].*

Within the BPM lifecycle, a business process is first described by a number of specifications (Definition 1) that describe individual sets of requirements. A requirement is defined as follows:

Definition 3 (Requirement). *Provision that contains criteria to be fulfilled [17].*

These individual sets of requirements together define the system requirements. The system requirements are depicted as the gray area in Fig. 1 and include the user requirements, design properties, and regulations. The system requirements are defined as follows:

Definition 4 (System requirements). *A structured collection of requirements— comprising functions, performance, design constraints, and attributes—of the system and its operational environment and external interfaces [17].*

The system requirements can consist of different sets of specifications, including (i) the user requirements, (ii) the design properties of the chosen modeling method, and (iii) the regulations as imposed by external authorities. The user requirements, design properties, and regulations are defined as follows:

Definition 5 (User requirements). *The requirements for use that provide the basis for the design and evaluation of interactive systems to meet identified user needs [17].*

Definition 6 (Design properties). *The context-independent behavioral requirements of the created model given the chosen modelling method [2, 18].*

Definition 7 (Regulations). *Requirements, imposed by an authority, that establish the legal and illegal behaviors and states for a specific domain and jurisdiction [17].*

Given the system requirements (Definition 4), a model of the business process (Definition 2) can be derived through the process of refinement. Such business process models can describe the business process along a number of different perspectives, including the control flow, data, and resource perspectives. Moreover, business process models can be *descriptive* or *prescriptive*. A descriptive model describes the business process as it is performed in the real world, while a prescriptive model describes the business process as it should be performed [10]. The distinction is important since descriptive and prescriptive models fulfill very different roles during verification, roles that should be considered carefully. Another distinction can be made between *procedural* (or imperative) process models and *declarative* process models. Procedural process models use an imperative specification that describe step by step *how* a business process is performed, while declarative process models describe *what* is performed using, often, a logical representation. Note that declarative process models in many cases should be seen as declarative process *specifications* instead, while the actual model obtained from such a specification (sometimes also referred to as the declarative process model) is, in fact, imperative in nature. Although a process model is a specification in itself, the terms model and specification have distinct meaning in the area of verification and one should be careful when referring to logical representations as *models* when applying verification within the area of business process management. Sometimes, however, such a paradigm shift is correct, but should always be treated with extreme care. A business process model is defined as follows:

Definition 8 (Business process model). *A (graphical) representation of a business process that describes the typical business process instance in isolation by specifying the elements of the business process and their relationships along the control flow, data, and/or resource perspectives [5].*

Software systems can support business processes in many different ways. Given a business process model, software support may range from deployment of large information systems, such as business process management systems or case management systems, to individual software packages being used as each task is being performed manually in an ad-hoc way. We refer to the collection of hardware and software systems that support the business process as the implementation, which is defined as follows:

Definition 9 (Implementation). *Result of translating a design into hardware components, software components, or both, whose validity can be subject to test [15, 17].*

These software systems record information observed during execution of each process instance, or case, of a business process in a so-called event log [3]. The information captured in such event logs can be used to not only discover, monitor, and improve processes as supported by the software systems, but also to verify their correct execution against the requirements and regulations.

Definition 10 (Event log). *A collection of traces, where each trace is an ordered sequence of events observed and recorded during the execution of an instance/case of a business process. Each event refers to an action performed by an actor or the supporting implementation at a particular time, for a particular case, and possibly includes relevant data concerning that case [3].*

2.2 Verification Relations

Given the process of verification, between the described artifacts two possible relations can be proven: (i) relations that establish *conformance*, and (ii) relations that establish *compliance*. The first defines a relation between a specification and an implementation, while the latter defines a relation between two specifications. More formally:

Definition 11 (Conformance). *A relation between a specification and an implementation that holds when (observed behavior of) the implementation fulfills all requirements of the specification (when the implementation conforms to the specification) [15, 20].*

Definition 12 (Compliance). *A relation between two specifications, A and B, that holds when specification A makes requirements which are all fulfilled by specification B (when B complies with A) [15].*

3   The Goals of Verification Within Business Process Management

Business processes are verified towards a number of different goals. Existing verification techniques can be classified into those that have the goal of system conformance, process conformance, model conformance, model compliance, or regulatory compliance. Note that the strict definition of compliance (Definition 12) describes a relation between two specifications and not a relation between a specification and implementation. As a result, the goals of system and process compliance are included under regulatory compliance. Each of these goals may have multiple supporting techniques. Such techniques have the same goal, but often use different artifacts at different stages of the BPM lifecycle. We discuss these goals and each related technique.

3.1   System Conformance

The verification of a system's implementation against its specification in a process model is referred to as *system conformance*. In this definition, the word conformance refers only to the conformance relation of Definition 11 and not to the collection of popular process mining techniques. In general, conformance is restricted to a limited set of requirements to check against particular aspects and elements, or so-called *conformance points*. Accordingly, the implementation is verified against said conformance points [20]. The technique is depicted in Fig. 1 as the arrow from model to implementation, and is defined as follows:

Definition 13 (System conformance checking). *The process of verifying conformance of the implementation towards the business process model.*

System conformance checking is possible when the implementation is fully supported and automatized by a workflow engine. This type of verification can be applied during different stages of the BPM lifecycle. During design time, the operation of checking can either be reduced to the formal verification of the implementation, or employs testing to ensure that the behavior of the implementation reflects the expected behavior described by the process model. During runtime, typically the event log is used as a proxy for the implementation. However, in general we cannot fully depend on event log data for this purpose, as some computations can produce the same result for some instances, but a model may require a particular type of implementation or calculation.

3.2 Process Conformance

When verifying the behavior of an implementation (as observed in e.g. an event log) against a process model, this is commonly referred to as *process conformance checking* and describes the collection of popular process mining techniques that are either applied online, at runtime, or after-the-fact. During runtime, there is no clear difference between system conformance checking and process conformance checking; in general, process conformance checking is a subcase of system conformance checking. The technique is depicted in Fig. 1 by the arrow from model to event log, and is defined as follows:

Definition 14 (Process conformance checking). *The process of verifying the conformance of the observed behavior of the implementation, as recorded in the event log, towards the business process model.*

In this definition, the word conformance may refer to both the conformance relation (Definition 11) and the collection of popular mining techniques. The specification is represented by a prescriptive normative process model that describes the intended behavior based on best practices, business rules, company policies, legal requirements, etc. The event log is again used as a proxy for the implementation, which implies that the conformance points are limited to the tasks in the process and their contents. Process conformance checking verifies whether the actual behavior of the system matches the prescribed behavior of the normative model, identifies (un)common behavior and new behavior that is not specified or allowed in the model, and reports on deviations.

One of the central concepts in process conformance checking is a so-called *alignment*, which describes a relation between a trace and an execution of a process model as a sequence of moves, relating events in the event log to activities in the model [4, 10]. The moves in an alignment can be either a move on log, a move on model, or a synchronous move. An asynchronous move (i.e. a move on log or a move on model) incurs a cost, so that the *optimal* alignment (i.e. the closest match possible between the event log and the model) is defined as the alignment with the lowest total cost.

Another well-known approach uses a unified model of concurrent behavior called *event structures* [11]. In this approach, the event log and process model are each converted into an event structure, which are subsequently aligned via an error-correcting synchronized product. This is specifically suitable in cases where compact context-dependent feedback is required on deviations between the event log and process model.

3.3 Model Conformance

Event logs can be used as a specification to determine whether the process model provides an accurate depiction of the actual behavior, process or implementation. The verification technique used is still conformance checking, but we will refer to it as 'conformance checking for repair' to highlight the difference. The technique is depicted in Fig. 1 as the arrow from event log to model, and is defined as follows:

Definition 15 (Conformance checking for repair). *The process of verifying the conformance of the normative behavior of the business process model towards the observed behavior of the implementation, as recorded in an event log.*

In this definition, the word conformance refers to that of the relation defined in Definition 11. The relation of the process described here to the term conformance checking (Definition 14) is also relevant, as it effectively reverses the artifacts to be verified: the specification artifact is represented by the event log, whereas the model artifact is represented by the (descriptive) process model. That is, conformance checking for repair aims to identify scenarios where the model does not accurately describe the actual behavior as observed in the event log, to subsequently alter, or 'repair', the model by trying to incorporate the additional behavior observed from the event log. The idea is to alter the model such that it improves the correspondence between the model and the log as much as possible, usually by allowing inserting or skipping of activities. As such, the approach searches for models that are optimal in terms of fitness. That is, the fraction of behavior that is in the log but not possible according to the model is minimized.

Similar to process conformance, conformance for repair centralizes around the concept of alignment, where alternatives are provided to amend the model that optimizes the alignment such that the event log fits the repaired model at least as well as it fits the original model (see e.g. [21]). Alternative approaches offer an incremental procedure, where differences between the model and the log are presented to the user, who can subsequently choose whether or not to repair the difference (see e.g. [7]).

3.4   Model Compliance

Business processes are generally modeled following a certain standard such as the Business Process Model and Notation (BPMN) standard [16]. Standards like BPMN specify the elements and relations between elements allowed within its specified graphical notation of a business process model, how each element behaves, and more. Consequently, the used standard directly influences the design properties (Definition 6) of the model. Model compliance aims to verify not only syntactic adherence of the business process model to the used standard, but also semantic adherence to the design properties.

Correctness checking is the technique that verifies whether a process model is compliant with its design properties, and includes well-known techniques such as workflow-net soundness [1]. Note here that the term soundness specifically applies to correctness properties of the Petri-net based workflow-nets and should only be used when an intermediate workflow-net representation of the business process is used when establishing correctness. The correctness technique is depicted in Fig. 1 by the arrow between the business process model and design properties artifacts, and is defined as follows:

Definition 16 (Correctness checking). *The process of verifying compliance of the business process model towards the design properties.*

When using this technique, the act of verification entails using the business process model as the model for verification and checking it against a specification that describes the design properties. In this definition, the word compliance refers directly to the compliance relation of Definition 12 and not to that of regulatory compliance, which is discussed in the next section.

3.5 Regulatory Compliance

Companies are subject to large numbers of regulations (Definition 7) that affect the way they do business. When asked by authorities, companies must be able to prove that they comply with regulations, or be prepared to face large fines. In other words, they must prove regulatory compliance:

Definition 17 (Regulatory compliance). *Doing what has been asked or ordered, as required by rule or law [17].*

Regulatory compliance of business processes can be proven at different stages of the BPM lifecycle, while using different artifacts. At each stage, different techniques are required to verify whether a process model, a running instance of a process, or a process log adheres to a set of relevant regulations. Here we specifically use the word adheres because the different techniques, applied at the different stages of the BPM lifecycle, define different types of relations, i.e., compliance or conformance (Definitions 11–12).

At design time, the implementation does not exist and there are no running instances that generate data. Therefore, all that can be done is to check whether the specification of the process model *complies* (Definition 12) with the specification stating the regulations. In doing so, the technique attempts to prove compliance not only from the control flow perspective, but also other perspectives using semantic annotations [22]. Although it is possible to fully prove compliance of certain sets of regulations at design time, in most cases this process should be considered a preventative measure that attempts to *mitigate the risk of violating the regulations*. To ensure anything further, one must also prove the process was actually followed when performed (e.g., by proving process conformance). Nevertheless, the technique has no access to data from runtime instances and, therefore, can often not cover the full set of regulations. The technique is depicted in Fig. 1 by the arrow from regulations to model, and is defined as follows:

Definition 18 (Regulatory compliance checking). *The process of verifying compliance of the business process model towards the regulations in order to prove or disprove regulatory compliance of the modelled behavior.*

At runtime, data from running process instances can be used to determine whether the enactment satisfies the conditions given by the regulations. The activity can be understood as a *conformance* relation (Definition 11) where the conformance check points fully cover the requirements mandated by the regulations. Even if the conformance points cover the legal requirements, it is only possible to determine breaches against the regulations based on the events observed till the time when *regulatory compliance* (Definition 17) is checked by proving the conformance relation (Definition 11). However, we cannot use conformance to check if the full instance will satisfy the legal requirements, since—for the activities that have not been executed—we can only rely on the specified business process model to prove the *compliance* relation (Definition 12) for the remaining possible execution paths. The technique is illustrated in Fig. 1 by the arrow from the regulations to the event log, and is defined as below. Note that the name of the defined activity refers to *regulatory compliance* (Definition 17) even though the activity defines a *conformance* relation (Definition 11). This observation lies at the core of the discussion in the remainder of this position paper, and will be explored in detail.

Definition 19 (Runtime regulatory compliance checking). *The process of verifying the conformance of the currently observed behavior, as recorded in the event log, towards the regulations in order to prove or disprove regulatory compliance of the currently observed behavior.*

After-the-fact regulatory compliance checking, known as auditing, has access to the full instance data and can, therefore, prove regulatory compliance in its entirety. Using only this approach, however, is a high risk endeavor that companies prefer to mitigate as much as possible, because—at this point—any violation of the regulations that has happened cannot be rolled back anymore. As a result, regulatory compliance verification should occur at multiple stages of the BPM lifecycle to both mitigate risks of violations and prove regulatory compliance. For auditing, we speak of a *conformance* relation (Definition 11) where the set of conformance points cover the legal requirements to prove *regulatory compliance* (Definition 17). The technique is illustrated in Fig. 1 by the arrow from the regulations to the event log, and is defined as follows:

Definition 20 (Auditing). *The process of verifying the conformance of the observed behavior towards the regulations in order to prove or disprove regulatory compliance.*

4   Overview of the Relations and Goals of Verification

Within the area of BPM, the term business process conformance is mostly referred to in the context of the popular mining technique, while the term business process compliance generally refers to the context of regulatory compliance. In the context of verification, however, conformance and compliance are defined in the contexts of their *relations* (i.e., Definitions 11 and 12). When comparing perspectives, the use of the conformance and compliance terms does not match, as the *relation* and the *goal* of verification are used interchangeably. To highlight this mismatch between the verification relations and their goals, Table 1 summarizes the verification techniques described in Sect. 3. The table lists each technique together with the stage of the lifecycle it is applied, the artifacts used as the model and specification (i.e., Definitions 2–10), the type of relation (i.e., Definitions 11 or 12), and the goal of verification (i.e., Sects. 3.1–3.5).

Table 1. Overview of verification techniques in the context of BPM.

Verification technique	Lifecycle stage	Model artifact	Specification artifact	Relation type	Verification goal
System conformance checking	Implement	Implementation	Prescriptive model	Conformance	System conformance
Conformance checking	Enact	Event log	Prescriptive model	Conformance	Process conformance
Conformance checking	Diagnose	Event log	Prescriptive model	Conformance	Process conformance
Conformance checking for repair	Diagnose	Descriptive model	Event log	Conformance	Model conformance
Correctness checking	Design	Model	Design properties	Compliance	Model compliance
Regulatory compliance checking	Design	Model	Regulations	Compliance	Regulatory compliance
Regulatory compliance checking	Enact	Event log	Regulations	Conformance	Regulatory compliance
Auditing	Diagnose	Event log	Regulations	Conformance	Regulatory compliance

From Table 1, it can be observed that, between all verification techniques, only two relations are compliance relations, and both of these techniques use the business process model as the model for verification. Secondly, out of the other six techniques that have a conformance relation, only four have a conformance related goal. Finally, although three different verification techniques have the goal of regulatory compliance, only one has an actual compliance relation, while the others have conformance relations.

Given these observations, it is clear that there exists a gray area between the use of the conformance and compliance keywords among the verification relations and goals. The main 'offenders' are the techniques of *regulatory compliance checking* during enactment and *auditing*. These techniques both define *conformance* relations with the goal of checking regulatory *compliance*. Both these techniques were naturally developed out of the realization that proving a compliance relation between two specifications (i.e., model and regulations) could only provide so many preventative guarantees, and that runtime data and temporal information is required for definitive and complete results. It is not that these techniques are at fault. They very much prove regulatory compliance while defining a conformance relation. The conformance relation does not, suddenly, become a compliance relation when one has the goal of verifying regulatory compliance, nor does the goal suddenly become verifying regulatory conformance.

Even though the compliance and conformance terms are effectively synonyms in everyday language, it remains especially important that both research and application have clearly defined lines between developed and applied techniques and their related *keywords*. In literature, however, the conformance and compliance keywords are increasingly used interchangeably, which may cause confusion around the positioning and application of the different verification techniques within the research community itself, as well as in their application areas.

To ameliorate the issue, we must establish clear boundaries for the use of the conformance and compliance keywords within the context of verification during the BPM lifecycle. Figure 2 illustrates a step towards our proposed solution, featuring an additional gray area compared to Fig. 1 that represents business process execution. It includes the subset of BPM lifecycle artifacts used and created during enactment.

Given the additional area, we can now see that we can define correct boundaries through the use of three keywords instead of two. These keywords are (i) compliance, (ii), conformance, and (iii) regulatory compliance. That is,

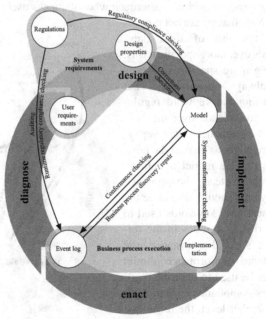

Fig. 2. Verification techniques applied during the BPM lifecycle (continued).

when we speak of *compliance*, we are applying verification using a specification from the system requirements and the business process model as the model for verification. On the other hand, when we speak of *conformance*, we are applying verification using the business process model with artifacts within the business process execution area. Finally, when we speak of *regulatory compliance*, we are applying verification using the regulations as the specification and artifacts within the business process execution area as the model for verification. Note that we use *compliance* (instead of regulatory compliance) to cover the verification of a model against regulations. Although this creates an area of overlap, this is not harmful since it correctly refers to compliance on both the relation and the regulatory goal. Moreover, when verifying (subsets of) the system requirements against a more refined set of such requirements, or a business process model against a more refined business process model, it is also *compliance*.

From this, it is clear that when using these three terms, it introduces clear boundaries that should be used to distinguish between verification techniques applied within the BPM lifecycle. The result is illustrated in Fig. 3, and should help both research and application to position work, accurately describe requirements, and interpret results. For example, consider an approach that obtains a business process model from an event log using a process mining technique and checks system requirements (e.g., regulations or user requirements) against the obtained model. That is, it obtains a model that describes the business process as it is performed in the real world (i.e., a descriptive model) from observed behavior of the implementation, and checks it against a specification. In this case, the approach would be a *regulatory compliance* approach when it verifies against regulations, a *compliance* approach when it verifies against design properties, and a *requirements validation* approach when it checks user requirements.

Note that we are not proposing the use of these keywords over more specific terms. Using more specific keywords is always encouraged. That is, using the keyword regulatory compliance over the keyword compliance when verifying regulations against the business process model is entirely correct. Instead, the proposed keywords should always be the highest level keywords used to describe techniques in the relevant areas. For instance, the keyword conformance should never be used to describe regulatory compliance even though, at a higher level, the technique describes a conformance relation. By following these guide-

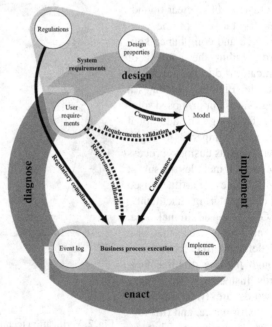

Fig. 3. Business process conformance and compliance during the BPM lifecycle.

lines, the community is ensured of using non-conflicting terminology and the proper positioning and application of techniques.

5 Discussion

The definition of clear boundaries between available techniques and tools is important for both researchers and practitioners. For researchers, it is not only important to ensure that the right terminology is used when describing their techniques and tools, but also to assist practitioners to select the correct tool for its intended purpose. Furthermore, such boundaries allow researchers to properly position their work, including the use of examples, selection of relevant related work, and evaluating against relevant work. For practitioners, on the other hand, it is important to ensure the validity of the results. That is, to ensure that the applied technique or tool verifies what was intended to be verified and be able to rely on the results and draw correct conclusions from those results. Consequently, more precise terminology allows to select the right portfolio of tools to collectively verify each aspect of the design and its implementation against each aspect of the set of system requirements, including user requirements, design properties, and regulations.

The question, however, remains whether some of the discussed techniques are possibly of value to the goals set for the other techniques. That is, we must discuss whether we actually should make the proposed distinction, or whether this is merely an intellectual issue. To do so, we discuss the relevance of some techniques to the goals set for the other techniques. That is, we discuss whether the technique of process conformance checking (Definition 14) is relevant to the goal of regulatory compliance (Sect. 3.5). Similarly, we discuss whether the technique of regulatory compliance checking (Definition 18) is relevant to the goal of process conformance (Sect. 3.2), and finally, we discuss whether the technique of process conformance checking (Definition 14) is always relevant to conformance from a legal point of view. In the remainder of this section, we discuss these questions, highlight any advantages or limitations that such applications yield, and present any analysis gaps that such applications may permit.

5.1 Should Process Conformance Be Used to Prove Regulatory Compliance?

As the popularity of process mining increased within the community, the idea slowly evolved that proving a *conformance* relation between an event log and a business process model can prove *regulatory compliance*. As such, the use of conformance checking techniques has been suggested as valuable to, for instance, agile compliance management [10] and GDPR [9]. Although technically conformance checking can be applied to prove regulatory compliance, it should be made clear that this approach is not ideal and can only prove regulatory compliance up to some point.

When using this approach, several strict conditions must be met, while results often lead to non-obvious inconclusive outcomes. First, a prescriptive business process model is required to check conformance. Second, this prescriptive model must be proven regulatory compliant using design time regulatory compliance checking (Definition 18).

One should be careful to note that, although design time regulatory compliance checking can check prescriptive models, it generally uses descriptive models. Third, the conformance checking must report any unfitting behavior. We must stress here that any unfitting behavior is not necessarily a violation of regulations. It simply means that a deviation was made from the possible executions described by the prescriptive model. As a result, this type of checking effectively denies any form of process flexibility.

Therefore, regulatory compliance can be proven through conformance checking by proving there is no unfitting behavior. However, it cannot prove that any unfitting behavior is an actual violation of regulations. One would still require additional regulatory compliance checking or auditing to prove this. In addition, it can only prove regulatory compliance along the control flow perspective, because the design time regulatory compliance checking techniques used to check the prescriptive model only has access to design time information and lacks process enactment information, such as data, resources, multiple instances etc. In this way, the limitations of the preventative measure of design time regulatory compliance checking (Definition 18) is transferred to an approach that in fact has process enactment information.

Although further model annotations of regulations are possible to consider other perspectives than that of the control flow, these approaches edge more towards also doing regulatory compliance checking while conformance checking, than just conformance checking—and would still deny any process flexibility. On the other hand, conformance checking approaches that enable process flexibility by allowing a certain level of unfitting behavior can never prove regulatory compliance without applying some form of actual regulatory compliance checking. As a result, the approach of using conformance to check regulatory compliance will always remain sub-optimal and should ideally be avoided. However, by continuing to use the keywords of conformance and compliance interchangeably, or using regulatory compliance examples to position conformance work, this approach may become common within application areas despite its non-ideal application.

5.2 Should Regulatory Compliance Be Used to Prove Process Conformance?

The application of regulatory compliance (Definition 18) to prove process conformance may, at first sight, seem completely irrelevant. However, it is possible but requires an unconventional approach. Again, it should be made clear that this approach is not ideal and can only prove conformance up to some point. That is, the approach can only obtain a degree of fitness and not a degree of precision. To obtain a degree of fitness of an event log with respect to a process model using regulatory compliance, we must first obtain a declarative specification of the prescriptive business process model. That is, we must obtain a set of declarative rules (e.g., temporal logic expressions) that together describe all possible paths within the business process model.

One example to automatically obtain such a declarative specification includes obtaining an event structure from (sets of) process model(s) and extracting a specification in the form of computation tree logic expressions [8]. Once a declarative specification is obtained, execution traces of the business process (captured by the event log) can be evaluated against the declarative specification using formal regulatory compliance verification techniques such as existing model checking tools and packages [12, 13].

To obtain a degree of fitness for an execution trace, or all execution traces within the event log, we can divide the number of satisfied temporal logic expressions by the total number of temporal logic expressions being verified. In this way, the degree of fitness decreases as more temporal logic expressions are violated.

Next to the degree of fitness, results include sets of satisfied and violated temporal logic expressions. Consequently, these results will be difficult to interpret by non-experts. As a result, the approach to use regulatory compliance to check conformance is non-ideal due to partial and difficult to interpret results, and should be avoided. By continuing to use the keywords of conformance and compliance as being interchangeable, or using regulatory compliance examples to position conformance work, this approach may, however, appear within application areas despite its non-ideal application.

5.3   Should Process Conformance Always Be Used to Prove Legal Conformance?

In a previous section, we gave a short outline how to use what we called process conformance to prove regulatory compliance from the process oriented information systems point of view. In this section, we are going to look at the issue from a legal point of view. First of all, in legal documents there is often no real distinction between compliance and conformance (and, sometimes the two english terms are translated to a single term in other languages). The two terms both generically mean to obey to a set of prescriptions. For instance, consider the proposal for the European Union's Artificial Intelligence (AI) Act. According to the current proposal, AI (and more generally) systems operating in specific sectors have to comply with the Act, as the explanatory text recites:

> "Those AI systems will have to *comply* with a set of horizontal mandatory requirements for trustworthy AI and follow *conformity* assessment procedures before those systems can be placed on the Union market."

As we can see, the Act does not differentiate between the model of an AI system and its implementation. Furthermore, the Act seems to indicate that compliance refers to the behavior of day-to-day operations of the implementation; on the contrary, systems have to obtain conformity certificates before the system is placed on the market or operates in the European Union. Accordingly, conformance certificates are based on the evaluation of the systems before the systems are deployed. This poses the question if process and system conformance as understood in the business process community (as discussed in the previous sections) offer suitable techniques for providing conformance certificates for AI systems against the requirements set by the Act. The answer seems to be negative, since the requirements for conformance certificates appears to be closer to what we called regulatory compliance. Thus, while some of the techniques and methodologies developed for business processes appear adequate for the AI Act, the terminology used to describe them might not correspond to the terminology used by the legal and business communities; therefore, there is risk that BPM solutions will not fit for some applications or arc evaluated with negative results, and effective techniques not to be adopted, limiting the impact of BPM technology for this important market.

6 Conclusion

The notions of conformance and compliance received substantial attention in the past decade in the BPM community. Often the two terms are used interchangeably, both in the field and in the broader community. However, from a technical point of view, they have been proposed with a different meaning. In general, compliance and conformance are two types of verification of systems, relating two BPM artifacts. In this paper, we provided comprehensive definitions of the two notions and activities related to them throughout the lifecycle of the development and deployment of process aware information systems and the artifacts they apply to (i.e., design specifications and regulatory frameworks, process models, implementations, and event log). While there are surface similarities among the verification problems and the activities specific to one of them, we discuss some of the reasons why, in general, effective methods for one particular type of verification (e.g., conformance) cannot guarantee a successful verification for a different type of relation (e.g., compliance). Accordingly, the discussion pointed out the need for a uniform set of definitions (and this is what we attempted in this contribution), and consequently, a unified terminology to present them. Finally, we addressed the problem whether the notions used in the BPM community have a counterpart in the wider audience, in particular, in the legal domain, where the terms are often used. It turns out that the picture is not so clear, given that the notions are used with their commonly understood meaning (corresponding essentially to what we call regulatory compliance) and not with their technical meaning. The major observation is that when interacting with external partners, first one has to understand what is the verification problem to be addressed, and then to determine what are the technical capabilities to be used. We believe that the discussions about the different techniques (and the shortcomings of using other techniques) offer guidelines to see how to succeed in the tasks based on BPM technology.

References

1. Aalst, W.M.P.: Verification of workflow nets. In: Azéma, P., Balbo, G. (eds.) ICATPN 1997. LNCS, vol. 1248, pp. 407–426. Springer, Heidelberg (1997). https://doi.org/10.1007/3-540-63139-9_48
2. van der Aalst, W.M.P.: The application of petri nets to workflow management. J. Circuits, Syst. Comput. 8(01), 21–66 (1998)
3. van der Aalst, W.M.P.: Process Mining-Data Science in Action, 2nd edi. Springer, Heidelberg (2016). https://doi.org/10.1007/978-3-662-49851-4
4. van der Aalst, W.M.P., Adriansyah, A., van Dongen, B.: Replaying history on process models for conformance checking and performance analysis. Wiley Interdisc. Rev. Data Min. Knowl. Discov. 2(2), 182–192 (2012)
5. van der Aalst, W.M.P., Weijters, T., Maruster, L.: Workflow mining: discovering process models from event logs. IEEE Trans. Knowl. Data Eng. 16(9), 1128–1142 (2004)
6. van der Aalst, W.M.P., ter Hofstede, A.H.M., Weske, M.: Business process management: a survey. In: van der Aalst, W.M.P., Weske, M. (eds.) BPM 2003. LNCS, vol. 2678, pp. 1–12. Springer, Heidelberg (2003). https://doi.org/10.1007/3-540-44895-0_1

7. Armas Cervantes, A., van Beest, N.R.T.P., La Rosa, M., Dumas, M., García-Bañuelos, L.: Interactive and incremental business process model repair. In: Panetto, H., et al. (eds.) OTM 2017. LNCS, vol. 10573, pp. 53–74. Springer, Cham (2017). https://doi.org/10.1007/978-3-319-69462-7_5

8. van Beest, N.R.T.P., Groefsema, H., García-Bañuelos, L., Aiello, M.: Variability in business processes: automatically obtaining a generic specification. Inf. Syst. **80**, 36–55 (2019)

9. Burattin, A., van Zelst, S.J., Armas-Cervantes, A., van Dongen, B.F., Carmona, J.: Online conformance checking using behavioural patterns. In: Weske, M., Montali, M., Weber, I., vom Brocke, J. (eds.) BPM 2018. LNCS, vol. 11080, pp. 250–267. Springer, Cham (2018). https://doi.org/10.1007/978-3-319-98648-7_15

10. Carmona, J., van Dongen, B.F., Solti, A., Weidlich, M.: Conformance Checking-Relating Processes and Models. Springer, Cham (2018). https://doi.org/10.1007/978-3-319-99414-7

11. García-Bañuelos, L., van Beest, N.R.T.P., Dumas, M., La Rosa, M., Mertens, W.: Complete and interpretable conformance checking of business processes. IEEE Trans. Software Eng. **44**(3), 262–290 (2017)

12. Groefsema, H., van Beest, N.R.T.P., Aiello, M.: A formal model for compliance verification of service compositions. IEEE Trans. Serv. Comput. **11**(3), 466–479 (2018)

13. Groefsema, H., van Beest, N.R.T.P., Armas-Cervantes, A.: Automated compliance verification of business processes in apromore. In: BPM Demo Track (CEUR), vol. 1920, pp. 1–5 (2017)

14. Groefsema, H., Bucur, D.: A survey of formal business process verification: from soundness to variability. In: Third International Symposium on Business Modeling and Software Design (BMSD 2013), pp. 198–203. SciTePress (2013)

15. International Organization for Standardization: Information technology – open distributed processing, reference model: Overview part 1. Standard ISO/IEC 10746–1:1998, International Organization for Standardization, Geneva, CH, December 1998. https://www.iso.org/standard/20696.html

16. International Organization for Standardization: Information technology – object management group business process model and notation. Standard ISO/IEC 19510:2013, International Organization for Standardization, Geneva, CH, July 2013. https://www.iso.org/standard/62652.html

17. International Organization for Standardization: Systems and software engineering – vocabulary. Standard ISO/IEC/IEEE 24765:2017(E), International Organization for Standardization, Geneva, CH, September 2017. https://www.iso.org/standard/71952.html

18. Kiepuszewski, B., ter Hofstede, A.H.M., Bussler, C.J.: On structured workflow modelling. In: Wangler, B., Bergman, L. (eds.) CAiSE 2000. LNCS, vol. 1789, pp. 431–445. Springer, Heidelberg (2000). https://doi.org/10.1007/3-540-45140-4_29

19. Ko, R.K.L.: A computer scientist's introductory guide to business process management (BPM). Crossroads **15**(4), 4:11-4:18 (2009)

20. Milosevic, Z., Bond, A.: Digital health interoperability frameworks: Use of RM-ODP standards. In: EDOC Workshop 2016, pp. 1–10 (2016)

21. Polyvyanyy, A., van der Aalst, W.M.P., ter Hofstede, A.H.M., Wynn, M.T.: Impact-driven process model repair. ACM Trans. Software Eng. Methodol. **25**(4), 1–60 (2016)

22. Sadiq, S., Governatori, G., Namiri, K.: Modeling control objectives for business process compliance. In: Alonso, G., Dadam, P., Rosemann, M. (eds.) BPM 2007. LNCS, vol. 4714, pp. 149–164. Springer, Heidelberg (2007). https://doi.org/10.1007/978 3 540-75183-0_12

A Data-Centric Approach to Design Resilient-Aware Process Models in BPMN

Simone Agostinelli, Francesca De Luzi, Umberto di Canito, Jacopo Ferraro, Andrea Marrella[✉], and Massimo Mecella

Sapienza Università di Roma, Rome, Italy
{simone.agostinelli,francesca.deluzi,umberto.dicanito,jacopo.ferraro,
andrea.marrella,massimo.mecella}@uniroma1.it

Abstract. The widespread diffusion of Internet-of-Things (IoT) technologies is prompting organizations to rethink their business processes (BPs) towards incorporating the data collected from IoT devices directly into BP models for improved effectiveness and timely decision making. Nonetheless, IoT devices are prone to failure due to their limitations in terms of computational power and energy autonomy, leading to compromise the availability and quality of the collected data, with the risk to prevent the correct execution of the entire BP. To mitigate this issue, resilience is a feature that any data-aware BP should support at design-time, by focusing on the role of available - as an alternative to unreliable - data as a resource for increasing BP robustness to failures. In this paper, we formalize an approach for designing and evaluating resilient-aware BP models in BPMN (Business Process Modeling and Notation) through a maturity model that takes into account their degree of awareness through levels of resilience, which can be computed using the provided formalization. In addition, we show how to extend the metamodel of BPMN 2.0 to address the proposed resiliency levels, and we investigate the feasibility of the approach through a user evaluation.

1 Introduction

With the widespread diffusion of Internet-of-Things (IoT) technologies and the exponential growth of generated data, it is becoming crucial for organizations to rethink their business processes (BPs) towards incorporating the data collected from IoT devices directly into BP models for improved effectiveness and data-driven decision-making [8]. For instance, in the logistics domain, IoT devices provide real-time monitoring of goods transportation in terms of their position or state (e.g., temperature, humidity, etc.), enabling the underlying BPs to optimize their operational efficiency. Nonetheless, when a BP becomes *data-aware*, there are also some side effects in terms of BP reliability. Since IoT devices are prone to failure due to their limitations in terms of computational power and energy autonomy, the risk exists that they might deliver data of low quality or stop working without any previous notice [10], preventing the correct BP execution.

In this context, a proper design of *resilient* BPs becomes fundamental. Resilience concerns the *"ability of a system to cope with unplanned situations in*

© Springer Nature Switzerland AG 2022
C. Di Ciccio et al. (Eds.): BPM 2022, LNBIP 458, pp. 38–54, 2022.
https://doi.org/10.1007/978-3-031-16171-1_3

order to keep carrying out its mission" [3]. Satisfying resilience requirements has been often considered as a run-time issue. According to [9,14], many approaches have been proposed to keep BPs running even when some unplanned exceptions occur at run-time, by implementing ad-hoc countermeasures during the execution stage of the BP life-cycle. However, this requires to know precisely where potential mistakes can manifest in the BP. This information, if not explicitly documented in the BP model, may lead to a defective implementation of compensatory strategies for such mistakes. As BP models can explicitly mark and indicate data elements involved in the BP, we can pinpoint the resiliency issues that BP might suffer directly at design-time. This means a shift of focus from *what to do* in case of failures to *what may be affected* when a failure occurs.

The goal of this paper is to provide an approach for designing and evaluating resilient-aware BP models where data are considered as "first class citizens", by driving the improvement of resilience to reduce the possible impact of failures caused by missing/unreliable data due to improper human behavior and/or IoT device errors. Specifically, we introduce a rigorous formalization of the approach that is based on assessing at design-time how available data re-definitions can possibly be exploited to design viable alternatives in the BP model to make it more resilient at run-time. In this direction, a maturity model for resilience awareness is proposed, based on a modeling notation extending BPMN (ISO/IEC 19510:2013 - Business Process Modeling and Notation). The maturity model is organized in five resiliency levels, which can be computed using the provided formalization and allow BP designers to model at an increasing degree of detail how data should be defined to have resilient by-design BP models. In addition, to capture the novel resiliency constructs introduced by our approach, we propose an extension to the BPMN 2.0 metamodel [12] that was exploited to develop a tool, called RES-BPMN, implementing our approach. Finally, we present the results of a user evaluation performed to study the feasibility of the approach.

The rest of the paper is organized as follows. After a discussion of the related work in Sect. 2, in Sect. 3 we introduce the main concepts of the BPMN notation and we present a motivating running example. Section 4 specifies the proposed maturity model and the resiliency levels. In Sect. 5, we show how to extend the metamodel of BPMN to address the resiliency levels. Finally, in Sect. 6, we investigate the feasibility of the approach and provide a critical discussion about its general applicability, by tracing future work.

2 Related Work

Resilience engineering has its roots in the study of safety-critical systems [6], which aim at ensuring that organizations operating in turbulent settings attain high levels of safety despite a multitude of emerging risks and complex tasks. In the BPM (Business Process Management) field, the concept of resilience has been mainly tackled through the notions of BP *flexibility* [14] and *risk-aware BPM* [20]. Research on BP flexibility has focused on four major needs to make BPs robust to business changes, namely *(i) variability* [15], *(ii) looseness* [1], *(iii) adaptation* [9], and *(iv) evolution* [4]. However, the ability to deal with changes makes BP flexibility a required, but not sufficient, means for building resilient BPs. While BP flexibility produces "reactive" approaches that deal with exceptions at design-time by incorporating remedial strategies into the BP model, or at run-time if any "known" disturbance arises, BP resilience requires "proactive" techniques accepting and managing change "on-the-fly" rather than anticipating it, to enable a BP to address new emerging and unforeseeable changes with the potential to cascade [11]. On the other hand, while relatively close to the concept of risk-aware BPM, which evaluates operational risks on the basis of historical threat probabilities, resilient BPM shifts attention to the "realized risks" and their consequences, to improve risk prevention and mitigation.

The amount of research works directly addressing BP resilience is quite limited. Among the most relevant, the work of Antunes [2] focuses on developing a set of services integrating resilience support in BPM systems, including detection, diagnosis, recovery and escalation. The approach of Zahoransky [23] investigates the use of process mining to create probability distributions on the time behavior of BPs, which are used as indicators to monitor the resiliency level at run-time and indicate countermeasures if the level drops. The work [22] provides a framework and a set of measures based on the analysis of previous BP executions to evaluate BP resilience. Finally, in our previous work [13], we developed a conceptual approach coupled with a maturity model to build multi-party declarative BPs using OMG CMMN (Case Management Model and Notation).

If compared with the aforementioned papers, in this paper we rigorously formalize a maturity model through BPMN to build resilient-aware BP models at design-time by focusing on the reliability of data exchanged within the BP, which is an aspect neglected in the literature. This makes our approach specifically targeted to those BPs that require data awareness for their execution. While data-aware BPM is a highly debated topic in the BPM literature (see [17] for a summary), and it is considered as a major requirement to integrate BPM with IoT technologies [8], here we do not develop a new approach to integrate data into BP models. Conversely, we exploit (and slightly extend) the data features available in BPMN to handle generic BP descriptions that could be immediately implemented via customary BPMN technologies. In a nutshell, our target is to provide a means for evaluating in advance the impact of data-driven disturbances on the BP and improving BP resilience to failures.

3   Business Process Modeling Notation

BPMN provides a standard graphical notation for BP modelling, with an emphasis on control flow. It essentially defines a flowchart incorporating a range of diverse components, including *activity nodes*, denoting business events or items of work performed by humans or software applications, and *control nodes* capturing the flow of control between activities. Activity nodes and control nodes can be connected by means of a flow relation in almost arbitrary ways. BPMN also enables to represent the information flowing through the BP, such as documents,

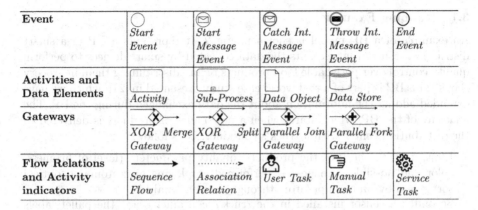

Event					
	Start Event	Start Message Event	Catch Int. Message Event	Throw Int. Message Event	End Event
Activities and Data Elements	Activity	Sub-Process	Data Object	Data Store	
Gateways	XOR Merge Gateway	XOR Split Gateway	Parallel Join Gateway	Parallel Fork Gateway	
Flow Relations and Activity indicators	Sequence Flow	Association Relation	User Task	Manual Task	Service Task

Fig. 1. A core subset of BPMN modeling elements

e-mails and other objects that are read or updated by means of dedicated *data elements*. As shown in Fig. 1, we take into account a (large) subset of BPMN elements including the data and control flow components considered in this paper. Hereafter, we describe the syntax of a BP model defined with such components.

Definition 1 (BP model). *A BP model specified in BPMN is a tuple* $\mathcal{N} = \langle \mathcal{O}, \mathcal{A}, \mathcal{G}, \mathcal{E}, \mathcal{F}, \mathcal{C}, Cond, \mathcal{D}, \mathcal{T}_{IN}, \mathcal{T}_{OUT} \rangle$, *where:*

- $\mathcal{O}$ *is a set of flow objects, which can be partitioned into disjoint sets of activities* $\mathcal{A}$, *gateways* $\mathcal{G}$ *and events* $\mathcal{E}$;
- $\mathcal{A}$ *is a set of activities, which can be atomic (i.e., tasks) or sub-processes;*
- $\mathcal{G}$ *is a set of gateways, which can be partitioned into disjoint sets of parallel gateways* $\mathcal{G}_P$ *for creating/synchronizing concurrent sequence flows, and XOR decision gateways* $\mathcal{G}_R$ *for selecting/joining a set of mutually exclusive alternative sequence flows based on data-driven conditions;*
- $\mathcal{E}$ *is a set of events, which can be partitioned into disjoint sets of start events* $\mathcal{E}_s$, *throw intermediate events* $\mathcal{E}_i^t$ *(e.g., a message that is sent) or catch intermediate events* $\mathcal{E}_i^c$ *(e.g., a message that arrives), and end events* $\mathcal{E}_e$;
- $\mathcal{F} \subseteq (\mathcal{O})$ x $(\mathcal{O})$ *is the sequence flow relation for connecting flow objects;*
- $\mathcal{C}$ *is a set of possible conditions that evaluate to true or false.*
- $Cond : \mathcal{F} \cap (\mathcal{G}_R$ x $\mathcal{O}) \rightarrow \mathcal{C}$ *is a function that maps sequence flows emanating from XOR decision gateways to conditions in* $\mathcal{C}$;
- $\mathcal{D}$ *is a set of data elements, which can be partitioned into disjoint sets of data objects* $\mathcal{D}_{ob}$ *(i.e., local data flowing through the BP) and data stores* $\mathcal{D}_{st}$ *(i.e., persistent databases that can be queried/updated by BP activities/events);*
- $\mathcal{T}_{IN} \subseteq (\mathcal{D}_{ob} \cup \mathcal{D}_{st})$ x $(\mathcal{A} \cup \mathcal{E}_e \cup \mathcal{E}_i^t)$ *is the input association relation used to link data elements to activities, end events or throw intermediate events.*
- $\mathcal{T}_{OUT} \subseteq (\mathcal{A} \cup \mathcal{E}_s \cup \mathcal{E}_i^c)$ x $(\mathcal{D}_{ob} \cup \mathcal{D}_{st})$ *is the output association relation used to link activities, start events or catch intermediate events to data elements;*

Without losing generality, we assume the behavior of BP models specified in BPMN to be ruled by the semantics described in [5].

3.1   Running Example

An example of a BP model is shown in Fig. 2. It represents a BP of a smart distribution centre that exploits the data collected by smart devices to perform quality control over perishable food products before distributing them in grocery shops. This BP is part of a real-world case study presented in [21], which we have extended adding the information about the data exchanged during the BP. The anatomy of the BP, which starts when a new pallet of products is delivered to the distribution center with a truck's container, is as follows:

- First, a quick check of the products' quality parameters (level of firmness, color and possible damages) is performed employing an automated optical sorter and by human operators through a visual analysis.
- Secondly, a sensor installed in the truck's container scans the pallet labels to obtain the products' information (e.g., product name, variety, collection date, etc.). Then, a second sensor captures the air temperature and humidity values related to the transport conditions. This information is recorded in a database and then used to evaluate the quality of the products.
- If the products' quality is considered as not adequate, the pallet is discarded. Conversely, if the quality of the products is good, the pallet is moved in the distribution centre and its storage is registered. The pallet is also temporally placed in a refrigerator room to prevent products' deterioration.
- At this point, a randomly selected sample of products is chosen from the pallet and analyzed in a laboratory to detect the presence of bacteria. If bacteria are detected, an alarm is triggered to indicate that the pallet must be discarded. Otherwise, the shipment procedure of the pallet starts.
- Finally, a last analysis is performed on the quality levels of the products in the pallet (e.g., to check if the firmness is optimal). If the quality is evaluated as not excellent, then the price of the products is dropped and the pallet is moved to a priority area to speed up its shipment and avoid further deterioration. When a truck is ready to start the distribution procedure, the pallet is loaded in a container for its shipment, and the BP completes.

By analyzing the BP behaviour, it is evident that the reliability of the data required to properly run the BP strongly depends on the reliability of the sensors employed for data collection. Any malfunctioning problem in sensors' behavior or connection issue will negatively impact the decision making and, consequently, the execution of the BP. According to [19], seven types of data flow anomalies can be detected in a BP: redundant data, lost data, missing data, mismatched data, inconsistent data, misdirected data, and insufficient data. We notice that all these anomalies can be classified into two main categories of issues related to the *availability* of data and their *quality* degree. In this direction, rather than automatically detecting structural data flow anomalies (e.g., like is investigated in [19]), we propose a maturity model that enables not only to uncover those data whose (un)availability and (low) quality can prevent the BP execution, but also suggests different countermeasures (weighted depending on the nature of the raised issues and the magnitude of their impact) to mitigate these negative effects and improve the BP resilience at design-time.

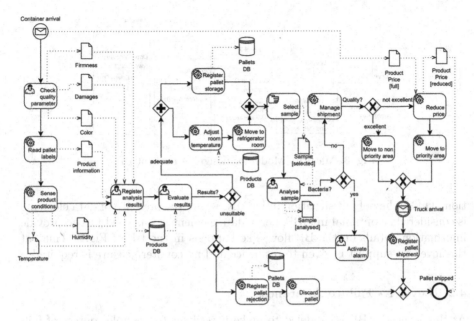

Fig. 2. BP model of the running example

4 Maturity Model

In this section, we present a maturity model with the aim to classify BPs modeled with BPMN in terms of their degree of resilience awareness. As shown in Fig. 3, the maturity model provides 5 levels of resilience awareness, which are defined on the basis of the ability of the BP to adjust itself to the possible unexpected failures with preparedness strategies to increase resilience at design-time. Starting from Level 0 (*No Resilience Awareness*), where resilience is not considered in the BP design, the other levels have been developed based on the three cornerstones of a resilient system as identified by [7]: Early detection (ED), Error tolerant design (ETD) and Recoverability (REC). Specifically, Level 1 (*Failure Awareness*) refers to ED, i.e., the recognition of system's weak signals that could be precursors of abnormal events. Level 2 (*Risk and Quality Awareness*) enforces ED by quantifying the impact of possible failures, and is the precondition for Level 3 (*Alternative Data Awareness*), which implements (ETD) by proposing alternative solutions that enable the system to still function well, but at reduced efficiency and marginally decreased quality. Finally, Level 4 (*Data Recovery Awareness*) refers to REC, which concerns the definition of recovery strategies to recover the system back to a normal state of operations.

4.1 Level 0 - No Resilience Awareness

At this level, a BP is modeled reflecting the desired scenario where it is assumed that all the data elements involved in the BP are available for its correct execu-

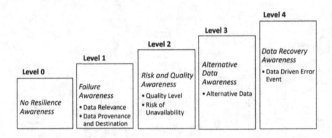

Fig. 3. Maturity Model for designing resilient BPs

tion. This is the default situation in BPMN, where the presence of data elements is considered as optional in a BP, i.e., a data element is supposed just to provide information details on the BP flow, like happens in the BP of Fig. 2. Thus, at this level, no support is given to resilience and no countermeasure is required.

4.2   Level 1 - Failure Awareness

At this level, the BP is modeled to make it resilient to possible sources of failure due to the *unavailability* of data elements, which might affect one or many activities that are consuming/producing such data. To have a clear map of which relevant data elements may be subject to failures, the BP designer is first required to identify them in the BP model and label them with the tag ⟨*true*, U, U⟩. The first tag parameter indicates that the data element will be considered *relevant* for BP execution, i.e., its unavailability may affect the execution of the BP flow objects to which it is connected. In the BPMN metamodel [12], this can be specified by turning the `DataState` parameter to *true* (see Sect. 5). If a data element becomes relevant, the flow objects that consume that data can not be executed until it becomes available. Similarly, a relevant data element produced by a flow object is checked for availability in output when the execution of the flow object completes. If the data element is not available, an error is thrown. In this paper, we will use the boolean function $State(d)$ that is *true* if a data element $d \in \mathcal{D}$ is relevant. The second and the third tag parameters indicate, respectively, the *quality level* and the *risk of unavailability* of the data element. Both are initially set to U (i.e., UNDEFINED) and have no impact at this level.

Once identified the relevant data elements, to make the BP model compliant with Level 1, the BP designer must first indicate the "provenance" and the "destination" of each relevant data object, i.e., which activity/start event/catch intermediate event produces the data object and which activity/end event/throw intermediate event consumes the data object. This can be done in BPMN exploiting the Association relation. Similarly, for each relevant data store, it must be specified at least a flow object that reads/updates data from/into it. Consequently, a *Level-1 compliant model* can be formally defined as follows:

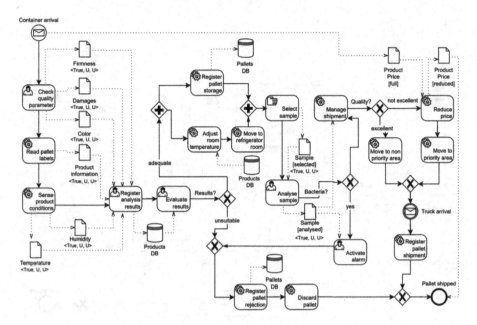

Fig. 4. Level 1 (Failure Awareness) compliant BP model

Definition 2 (Level-1 compliant model). *Let* $\mathcal{N} = \langle \mathcal{O}, \mathcal{A}, \mathcal{G}, \mathcal{E}, \mathcal{F}, \mathcal{C}, Cond,$ $\mathcal{D}, \mathcal{T}_{IN}, \mathcal{T}_{OUT} \rangle$ *be a BP model.* $\mathcal{N}$ *is said to be a "Level-1 compliant model" iff, for each* $d \in \mathcal{D}$ *with* $State(d) = true$ *one of the following conditions holds:*

- *$d \in \mathcal{D}_{ob}$, and there exist $x \in (\mathcal{A} \cup \mathcal{E}_s \cup \mathcal{E}_i^c)$, $y \in (\mathcal{A} \cup \mathcal{E}_e \cup \mathcal{E}_i^t)$, $t_i \in \mathcal{T}_{IN}$ and $t_o \in \mathcal{T}_{OUT}$ such that $t_i = \langle d, y \rangle$ and $t_o = \langle x, d \rangle$.*
- *$d \in \mathcal{D}_{st}$ and there exist $y \in (\mathcal{A} \cup \mathcal{E}_e \cup \mathcal{E}_i^t)$ and $t_i \in \mathcal{T}_{IN}$ such that $t_i = \langle d, y \rangle$, or $x \in (\mathcal{A} \cup \mathcal{E}_s \cup \mathcal{E}_i^c)$ and $t_o \in \mathcal{T}_{OUT}$ such that $t_o = \langle x, d \rangle$.*

Let us consider the BP of the running example. To increase the resiliency level of the model we should set as relevant all those data whose unavailability may lead to possible failures, i.e., the data collected by smart devices (e.g., *Firmness, Humidity, Temperature,* etc.) or obtained after a visual/automated analysis performed by human operators (e.g., *Damages, Sample [analyzed]*). Then, to make the BP fully compliant with Level 1, we must check that the relevant data objects are associated to their producer/consumer. Thus, we need to add an output association from the data object *Sample [analyzed]* to the activities *Activate Alarm* and *Manage Shipment,* as shown in Fig. 4. If this data object becomes unavailable or unreliable, the risk exists that the alarm is wrongly triggered or the shipment of products with bacteria is performed with severe effects.

4.3 Level 2 - Risk and Quality Awareness

While at Level 1 the BP designer declares which data elements are likely subject to failures, at Level 2 there is a first attempt to concretely quantify the *quality*

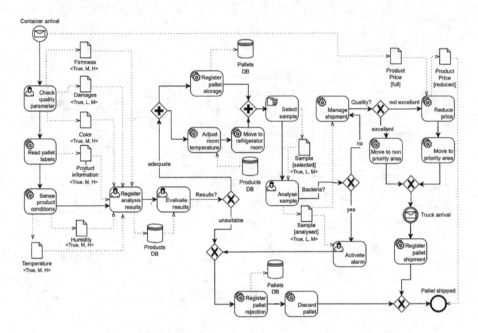

Fig. 5. Level 2 (Risk and Quality Awareness) compliant BP model

level and the *risk of unavailability* associated to such data elements. For the sake of simplicity, in the rest of the paper we assume the quality level/risk of a relevant data element bound to only four discrete values: U - UNDEFINED, L - LOW, M - MEDIUM, H - HIGH. The pair of parameters quality level/risk enables to build a kind of "criticality degree" that supports the BP designer to identify those data elements that might have more impact in case they are unavailable or have a poor quality. Note that, to switch from Level 1 to Level 2, it is required that, for any relevant data element, the quality level/risk are both assigned to a value different from U, i.e., they become objectively quantifiable. Of course, the choice of the values to associate to both parameters depends on the domain under observation. From a formal perspective, we introduce two functions $Quality(d)$ and $Risk(d)$ returning the quality level and the risk of unavailability of a relevant data element $d \in \mathcal{D}$, and we define a *Level-2 compliant model* as follows.

Definition 3 (Level-2 compliant model). *Let* $\mathcal{N} = \langle \mathcal{O}, \mathcal{A}, \mathcal{G}, \mathcal{E}, \mathcal{F}, \mathcal{C}, Cond,$ $\mathcal{D}, \mathcal{T}_{IN}, \mathcal{T}_{OUT} \rangle$ *be a Level-1 compliant model.* $\mathcal{N}$ *is said to be a "Level-2 compliant model" iff, for each* $d \in \mathcal{D}$ *with* $State(d) = true$, *then* $Quality(d) \neq U$ *and* $Risk(d) \neq U$.

In the case of our running example, many data objects are the results of activities performed automatically through the support of smart sensors supported by sophisticated software. For example, the first quality check involves the use of an optical sorter to measure the firmness of the products contained in the pallet and detect their color. Similarly, other sensors installed in the pallet

Element	Name	Annotator	Name
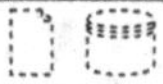 ⌐¬ ⌐⌐⌐⌐	*Alternative Data Element*	**X**	*No alternative for a data element*
Ⓓ	*Data Driven Error Event*	**R**	*Recoverable data element*

Fig. 6. Novel modeling elements and annotators

or in the truck container allow for a precise detection of products' information, temperature and humidity ((H)igh data quality). However, the electronic components of these devices are subject to deterioration due to their continuous usage, requiring scheduled/ad-hoc maintenance actions in case of malfunctioning ((M)edium risk of data unavailability). This means that data objects *Firmness, Color, Product Info, Temperature* and *Humidity* will be associated with the label $\langle true, M, H \rangle$. Conversely, to identify damaged products, a visual inspection is conducted, meaning a (potential) (M)edium quality level for the data object *Damages*. Similarly, the quality of *Sample [analyzed]* depends by the specific sample chosen, which leads to a (M)edium value for this parameter (cf. Fig. 5).

4.4 Level 3 - Alternative Data Awareness

Based on the information about the sources of failures and their potential impacts, the BP designer can decide to include alternative data in the BP model. Starting from the data elements with a higher risk of unavailability and lower data quality, the BP designer specifies if there are alternative data sources and how to reach them. To this aim, we introduce the function $Alt(d)$, which associates to a relevant data element $d \in \mathcal{D}$ an alternative data element $d_{al} \in \mathcal{D}$, or the special keyword 'X' if no alternative exists for d. This enables us to define data elements that act as *primary* data sources for some activities/events and others that work as their *alternatives*. As shown in Fig. 6 and in Fig. 7, we represent an alternative data element through a new BPMN icon with a shape identical to a "traditional" data element, but with a dashed border attached to the primary data source. If the BP designer is aware that no alternative is possible for a primary data, then the dashed border icon is labeled with 'X'.

Definition 4 (Level-3 compliant model). *Let* $\mathcal{N} = \langle \mathcal{O}, \mathcal{A}, \mathcal{G}, \mathcal{E}, \mathcal{F}, \mathcal{C}, Cond,$ $\mathcal{D}, \mathcal{T}_{IN}, \mathcal{T}_{OUT} \rangle$ *be a Level-2 compliant model.* $\mathcal{N}$ *is said to be a "Level-3 compliant model" iff, for each* $d \in \mathcal{D}$ *with* $State(d) = true$, *then: (i) there exists* $d_{al} \in \mathcal{D}$ *such that* $d_{al} \neq d$ *and* $Alt(d) = d_{al}$, *or (ii)* $Alt(d) = X$.

In our running example, we can associate the primary data objects having some risk of unavailability with a "backup" alternative version of the data. For example, if the optical sorter stops working, the human operators can employ a portable penetrometer to measure the products' firmness, and a spectrophotometer to perform color measurement based on spectral reflectance. Similarly, temperature and humidity can be obtained through portable temperature and

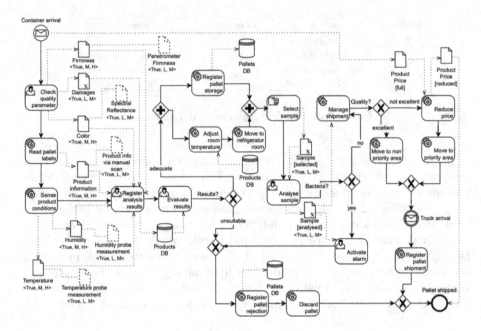

Fig. 7. Level-3 (Alternative Data Awareness) compliant BP model

humidity probes. Also the product information can be obtained employing a manual scanner. Of course, using manual devices to perform continuous measurements rather than automatic sensors can decrease the quality of the collected data. This means that the alternative data objects *Penetrometer Firmness, Spectral Reflectance, Product Info via manual scan, Temperature probe measurement* and *Humidity probe measurement* will be associated with the label $\langle true, L, M \rangle$. It is worth to notice that no alternatives exist for the data objects *Damages, Sample [selected]* and *Sample [Analyzed]*, i.e., the BP designer is declaring her awareness that these data represent single point of failures (cf. Fig. 7).

4.5   Level 4 - Data Recovery Awareness

In the previous level, we have discussed how the presence of alternative data allows us to substitute primary data sources if they are missing or unreliable. However, the quality of an alternative data is usually lower than its original counterpart, and sometimes this can be not adequate to progress with BP execution. To mitigate this issue, the final level of our maturity model pushes a BP designer to specify remedial actions to improve the quality of a data to a degree that is comparable to its original counterpart. These actions are triggered employing a new modeling element, named *data-driven error event*, which can be embedded in a event sub-process. In BPMN, event sub-processes are used to capture global BP exceptions and define recovery procedures. We represent a data-driven error event with a document marker within the event shape (see

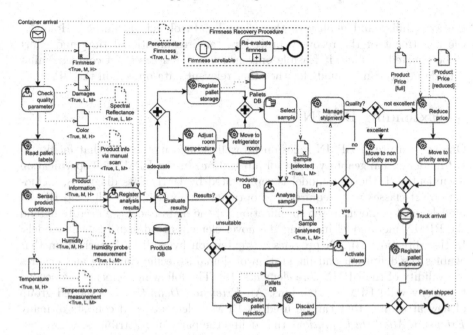

Fig. 8. Level-4 (Data Recovery Awareness) compliant BP model

Fig. 6). In our maturity model, we let the BP designer deciding if a data element requires to be restored trough a recovery procedure; if this is the case, then the icon of the data element to be recovered must be labeled with 'R'. At this point, to switch from Level 3 to Level 4, for any "recoverable" data element $d \in \mathcal{D}$, a data-driven error event $e_v \in \mathcal{E}_s$ is coupled with d and followed by a sub-process including the remedial actions to adjust its quality. From a formal perspective, we introduce the function $Rev(d)$ that associates to d a data-driven error event e_v, or the special keyword "NR" if d is considered as not recoverable.

Definition 5 (Level-4 compliant model). *Let $\mathcal{N} = \langle \mathcal{O}, \mathcal{A}, \mathcal{G}, \mathcal{E}, \mathcal{F}, \mathcal{C}, Cond, \mathcal{D}, \mathcal{T}_{IN}, \mathcal{T}_{OUT} \rangle$ be a Level-3 compliant model. $\mathcal{N}$ is said to be a "Level-4 compliant model" iff, for each $d \in \mathcal{D}$ with $State(d) = true$, then: (i) there exist a data-driven error event $e_v \in \mathcal{E}_s$, an end event $e_n \in \mathcal{E}_e$, a sub-process $a \in \mathcal{A}$, an event sub-process $a_{es} \in \mathcal{A}$, and two sequence flows f_1 and f_2 such that $Rev(d) = e_v$, $f_1 = (e_v, a)$, $f_2 = (a, e_n)$, and $\{e_v, f_1, a, f_2, e_n\} \in a_{es}$, or (ii) $Rev(d) = NR$.*

Concerning our running example, we can assume that if the optical sorter stops working and the amount of pallets to be checked is too high, then employing the portable penetrometer to measure the products' firmness becomes too time consuming for the human operators. Therefore, the BP designer can mark the data object "Firmness" with a 'R' and associate it to the data-driven error event called "Firmness unreliable". As shown in Fig. 8, this will trigger the starting of a recovery procedure that, for example, instructs to move the pallet in another area of the distribution center where an auxiliary optical sorter is located by restoring the availability and quality of the original data object "Firmness". However, the enactment of the recovery procedure requires additional time and effort to be enacted, making it feasible only in exceptional cases. Of course, similar considerations can be made for the other relevant data objects in the BP.

5   Extending BPMN

One key feature of BPMN relies on its well-defined metamodel that facilitates BP model exchangeability and tool integration. In the BPMN 2.0 specification document [12], the metamodel is represented by UML class diagrams, including object classes with required and optional attributes. Since all valid BPMN models must conform to the specifications of the metamodel, we need to extend the BPMN metamodel inserting the novel elements to design resilient models. In this direction, BPMN provides an "extension by addition" mechanism that enables the definition and integration of domain-specific concepts and ensures the validity of the BPMN core elements [18]. The following elements are needed to specify valid BPMN extensions. An *Extension Definition* is a named group of new attributes that can be used by BPMN elements, and consists of many *Extension Attribute Definitions* that define the particular attributes, whose values can be defined by the *Extension Attribute Value* class. To exploit the extension capabilities of BPMN, we have customized the well-known procedure for the methodical development of valid BPMN extensions provided by Stroppi in [18], which consists of the following steps (RES-BPMN is the name of our extension):

1. define a CDME (Conceptual Domain Model of the Extension) as UML class diagram that is able to capture the novel resiliency aspects;
2. define the RES-BPMN model based on the previous CDME model;
3. transform RES-BPMN into an XML Extension Definition Schema (EDS);
4. transform the XML EDS into an XML Schema Document.

Since our work mainly focuses on conceptual aspects and aims to create a maturity model, only the first two steps of the procedure are shown here. First, we identified a set of UML Class diagrams to be modified for capturing the novel BPMN elements (cf. Fig. 6): Data Object, Data Store, Data Association and Event. Then, for each of them, we created the CDME model, whose classes are typed as standard BPMN Concepts. Finally, the RES-BPMN model was derived by the application of the model transformation rules covering all possible CDME configurations to extend the existing Class Diagrams. For the sake of space, we focus here just on the extension of the Data Object Class Diagram (cf. Fig. 9). The complete list of CDME models and UML Class diagrams is available in an online appendix at: https://github.com/bpm-diag/RES-BPMN.

As shown in Fig. 9, we introduced new attributes to the BPMN standard, which are highlighted in bold. For failure awareness (Level 1), we exploit the existing optional DataState attribute, which indicates that the unavailability of a data object may affect the execution of the BP flow objects to which it is connected. By default, its value is set to *false*. For risk and quality awareness

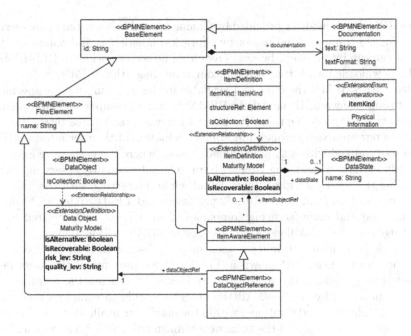

Fig. 9. RES-BPMN UML class diagram of data object class

(Level 2), we defined two attributes: `risk_lev` and `quality_lev`, which allow respectively to capture the unavailability risk and quality level of a data object through four values: U, L, M and H. By default, at Level 1, these attributes are set to U, i.e., their values are unknown a-priori. Alternative data awareness (Level 3) is captured with a boolean attribute `isAlternative`. In particular, for a given data object, `isAlternative` can be set either to *true* if an alternative version of the data exists (the association between a data and its alternative is made explicit throw a new class `DataAlternativeAssociation` created within the `DataAssociation` class) or *false*, i.e., there is no alternative for the data object. Finally, data recovery awareness (Level 4) is addressed by setting the attribute `isRecoverable` to *true*, which indicates that the BP designer can provide a recovery strategy for the data object through a data-driven error event. This is captured within the `Event` class.

6   User Evaluation and Concluding Remarks

Extending the metamodel of BPMN has allowed us to develop a software tool, called RES-BPMN, which implements our approach to systematically design resilient BP models in BPMN and check their compliance with the different levels of the maturity model. In the case of non compliance with a certain level, the tool suggests the steps to refine the BP model to achieve the desired level of resilience. RES-BPMN has been developed as an extension of bpmn.io, an open source BPMN web modeler provided by Camunda, and it is written in Javascript using NodeJS framework on top of two libraries: *diagram-js* and *bpmn-moddle*. Thus it can run into modern browsers requiring no server back-end. RES-BPMN can be downloaded at: https://github.com/bpm-diag/RES-BPMN.

Being RES-BPMN the only tool available in the literature for the specification of resilient-aware BP models in BPMN, no direct comparison was possible against other BP modeling tools. For this reason, we opted to investigate the feasibility of our approach through a usability evaluation of the user interface (UI) of the tool coupled with a thinking-aloud session, where the users were asked to explicitly execute a modeling task with an external evaluator observing them, indicating the methodological issues found while interacting with the UI. The users were selected from universities (2 professors and 4 PhD students), business (2 managers) and manufacturing companies (2 managers), and declared to be knowledgeable (60%), skilled (20%) or experts (20%) in BP modeling.

After a preliminary training session on introducing RES-BPMN, starting from the (not-resilient) BP shown in Fig. 2 and its description, the users were requested to systematically increase its resiliency level using the features and feedback provided by the tool. All the users were able to complete their task (providing different valid solutions) within the maximum available time (15 min). As soon as a user completed the task, we administered a SUS questionnaire [16]. SUS consists of 10 statements evaluated with a 5-point numerical scale that ranges from 1 ("strongly disagree") to 5 ("strongly agree"). At the end of the questionnaire, an overall score is assigned to it. We compared the score against the benchmark presented in [16], which associates to each range of the SUS score a percentile ranking varying from 0 to 100, indicating how well it compares to other 5,000 SUS observations performed in the literature. Since the obtained average SUS score was 80.8, according to the benchmark, the tool's usability corresponds to a rank of A, which indicates a degree of usability almost excellent.

We also collected valuable insights about the practical applicability of the approach during the thinking-aloud sessions. In particular, the users criticized the absence of an indicator to quantify the distance between a BP model and the complete achievement of a resiliency level. In this direction, as a future work, we plan to develop such an indicator exploiting our formalization of resiliency levels and measuring the number of modeling elements that are not compliant with the definitions in Sect. 4. In addition, by associating the quality level and the risk of unavailability of data elements with numeric weights, we can use them to build a quantifiable "criticality value" that identifies the data that might have more severe negative effects in case of their unavailability of low quality. This value could enrich the above indicator to provide a better understanding of the impact and the risks of a non-compliance with a resiliency level.

A second threat to the feasibility of the approach is about the practical conditions and assumptions under which it can be considered as effective. In particular, the users pointed out that the existence of alternatives might not be always guaranteed; analogously, resilience might also be affected by other factors different from data, like resource unavailability, temporal constraint violations, etc. In this paper, we focused on the data as main source of failures affecting BP resilience, and covering other potential factors is out of the scope of this work. However, the investigation of such factors is in the list of future works.

To sum up, we believe that measuring the usability of the UI of RES-BPMN is as a good preliminary indicator to validate the feasibility of our approach. The resiliency levels introduced in this paper, being based on a well-known standard such as BPMN, go in the direction of providing a reference framework for developing novel techniques and metrics to address BP resilience towards more accurate quantitative analysis. Of course, a general acceptance of the maturity model needs an extensive empirical evaluation of the approach.

Acknowledgments. This work has been supported by the H2020 project DataCloud and the Sapienza grant BPbots.

References

1. van der Aalst, W.M.P., Pesic, M., Schonenberg, H.: Declarative workflows: balancing between flexibility and support. Comput. Sci. R&D **23**(2), 99–113 (2009)
2. Antunes, P., Mourão, H.: Resilient business process management: framework and services. Expert Syst. Appl. **38**(2), 1241–1254 (2011)
3. Caralli, R.A., Allen, J.H., White, D.W.: CERT Resilience Management Model: A Maturity Model for Managing Operational Resilience. Addison-Wesley, Reading (2010)
4. Casati, F., Ceri, S., Pernici, B., Pozzi, G.: Workflow evolution. In: Thalheim, B. (ed.) ER 1996. LNCS, vol. 1157, pp. 438–455. Springer, Heidelberg (1996). https://doi.org/10.1007/BFb0019939
5. Dijkman, R.M., Dumas, M., Ouyang, C.: Semantics and analysis of business process models in BPMN. Inf. Software Technol. **50**(12), 1281–1294 (2008)
6. Hollnagel, E., Woods, D.D., Leveson, N.: Resilience Engineering: Concepts and Precepts. Ashgate Publishing Ltd., Aldershot (2007)
7. Jain, P., Pasman, H.J., Waldram, S., Pistikopoulos, E., Mannan, M.S.: Process Resilience Analysis Framework (PRAF): a systems approach for improved risk and safety management. J. Loss Prev. Proc. Ind. **53**, 61–73 (2018)
8. Janiesch, C., et al.: The Internet of Things Meets Business Process Management: A Manifesto. IEEE Syst. Man Cybern. Mag. **6**(4), 34–44 (2020)
9. Marrella, A., Mecella, M., Sardina, S.: Supporting adaptiveness of cyber-physical processes through action-based formalisms. AI Commun. **31**(1), 47–74 (2018)
10. Moore, S.J., Nugent, C.D., Zhang, S., Cleland, I.: IoT reliability: a review leading to 5 key research directions. CCF Trans. Pervasive Comput. Interact. **2**(3), 147–163 (2020)
11. Müller, G., Koslowski, T.G., Accorsi, R.: Resilience - A New Research Field in Business Information Systems? In: Abramowicz, W. (ed.) BIS 2013. LNBIP, vol. 160, pp. 3–14. Springer, Heidelberg (2013). https://doi.org/10.1007/978-3-642-41687-3_2
12. OMG: Business Process Modeling and Notation, Version 2.0.2, January 2014. http://www.omg.org/spec/BPMN/2.0.2/
13. Plebani, P., Marrella, A., Mecella, M., Mizmizi, M., Pernici, B.: Multi-party business process resilience by-design: a data-centric perspective. In: Dubois, E., Pohl, K. (eds.) CAiSE 2017. LNCS, vol. 10253, pp. 110–124. Springer, Cham (2017). https://doi.org/10.1007/978-3-319-59536-8_8

14. Reichert, M., Weber, B.: Enabling Flexibility in Process-Aware Information Systems - Challenges, Methods, Technologies. Springer, Heidelberg (2012). https://doi.org/10.1007/978-3-642-30409-5

15. La Rosa, M., van der Aalst, W.M.P., Dumas, M., Milani, F.: Business process variability modeling: a survey. ACM Comput. Surv. (CSUR) **50**(1), 1–45 (2017)

16. Sauro, J., Lewis, J.R.: Quantifying the User Experience: Practical Statistics for User Research. Morgan Kaufmann, Cambridge (2016)

17. Steinau, S., et al.: DALEC: a framework for the systematic evaluation of data-centric approaches to process management software. SOSYM **18**(4), 2679–2716 (2019)

18. Stroppi, L.J.R., Chiotti, O., Villarreal, P.D.: Extending BPMN 2.0: method and tool support. In: Dijkman, R., Hofstetter, J., Koehler, J. (eds.) BPMN 2011. LNBIP, vol. 95, pp. 59–73. Springer, Heidelberg (2011). https://doi.org/10.1007/978-3-642-25160-3_5

19. Sun, S.X., Zhao, J.L., Jr., Nunamaker, J.F., Sheng, O.R.L.: Formulating the Data-Flow Perspective for Business Process Management. Inf. Syst. Res. **17**(4), 374–391 (2006)

20. Suriadi, S., Weiß, B., et al.: Current Research in Risk-aware Business Process Management: Overview, Comparison, and Gap Analysis. CAIS **34**(1), 52 (2014)

21. Valderas, P., Torres, V., Serral, E.: Modelling and executing IoT-enhanced business processes through BPMN and microservices. J. Syst. Softw. **184**, 111139 (2022)

22. Zahoransky, R.M., Brenig, C., Koslowski, T.: Towards a Process-Centered Resilience Framework. In: ARES (2015)

23. Zahoransky, R.M., Koslowski, T., Accorsi, R.: Toward resilience assessment in business process architectures. In: Bondavalli, A., Ceccarelli, A., Ortmeier, F. (eds.) SAFECOMP 2014. LNCS, vol. 8696, pp. 360–370. Springer, Cham (2014). https://doi.org/10.1007/978-3-319-10557-4_39

The Early Process Catches the Weak Event: Process Latency and Strategies for Its Reduction

Anton Koelbel[1]([⊠]) [iD] and Michael Rosemann[2] [iD]

[1] Northern Institute of Technology Management, Hamburg, Germany
anton.koelbel@tuhh.de
[2] Centre for Future Enterprise, Queensland University of Technology, Brisbane, Australia
m.rosemann@qut.edu.au

Abstract. In an increasingly data-rich environment, new opportunities for the domain of Business Process Management are created based upon identifying, interpreting, and acting on new and earlier signals. This shifts the focus from process execution to process initiation. Process latency is defined as the time from occurrence of a need to the start of the respective fulfillment process. Based on a comprehensive literature review, secondary data from real-world case examples and international focus groups, this paper classifies three strategies to reduce process latency. For each of these three strategies, reactive and proactive approaches are differentiated. This classification, in its core, serves as a tool for structured ideation and thus, expands process designers' toolset for explorative BPM. Beyond this, the paper contributes by conceptualizing process latency as a novel process metric within the BPM discipline.

Keywords: Explorative BPM · Process latency · Event sensing · Event processing

1 Introduction

As in many industries the threat of disruption increases, it is no longer sufficient to rely on internal operational efficiency only for lasting success [1]. Organizations additionally need to develop resilience, i.e., the ability to withstand disruptive external changes [2]. The implied challenge of running the present while building the future is conceptualized in the so-called ambidextrous organization, fueled by the observation that exploiting operational efficiency and exploring new business opportunities in parallel has become a crucial success factor [3].

This development can also be observed in Business Process Management (BPM), the holistic management discipline concerned with the description and design of how work is performed in organizations [4, 5]. Traditionally, the exploitative perspective, which employs a problem-focused approach and aims at continuous improvements, has largely been the focus within BPM and respective tools have consequently reached a substantial maturity [1]. However, against the backdrop of the aforementioned amplifying relevance of the resilience and ambidexterity imperative, there have been recent calls

© Springer Nature Switzerland AG 2022
C. Di Ciccio et al. (Eds.): BPM 2022, LNBIP 458, pp. 55–69, 2022.
https://doi.org/10.1007/978-3-031-16171-1_4

for a more explicit consideration of an explorative perspective that supports the search for new value propositions [6, 7]. This search is characterized by the quest for what else is possible, a sharp contrast to the focus of exploitative BPM which concentrates on resolving identified pain points within a process.

In today's digital environment, it is especially a surge in the amount and accessibility of available data that catalyzes entirely new process design options [8, 9]. This data-intensity materializes in the context of BPM as manifestations of events (e.g., a door is closed, a parcel has been dropped, a mobile device entered a geofenced area) that were previously unnoticed. Each of these events comes with a potential signal that might trigger a request for a business process. It therefore becomes an opportunity to sense, interpret and act upon such external signals, if they are deemed relevant [10]. Organizations that master event sensing and subsequent processing can activate their processes earlier and by this gain an edge in terms of time-to-process, i.e., they benefit from low process latency [10].

This focus on and the increased opportunity to minimize process latency leads to two fundamental new perspectives in the context of BPM. First, early event identification, and not just process activities that can be eliminated, streamlined, and automated as part of lean management or RPA initiatives, is becoming an additional focus of process designers. Second, no longer is the process fulfillment time the only key temporal process metric. Process latency, defined as the time from occurrence of a need to the start of the respective fulfillment process, is becoming a relevant process metric in cases in which a fast, proactive process execution creates a 'first process advantage'. Therefore, the research question of this paper is: *How can process latency be reduced?*

In order to address this question, we developed a classification of process latency reduction strategies through a two-staged research methodology. First, we studied the extant literature on process redesign and latency reduction strategies to build an a-priori conceptualization. Second, we conducted a series of focus groups with global BPM experts from academia and industry to identify case examples so we could empirically assess its completeness, relevance and applicability and develop additional principles for operationalization.

The remainder of this paper is structured as follows: Hereinafter, Sect. 2 introduces the overall research context by providing a summary of the body of knowledge from relevant disciplines. In Sect. 3, we outline the research methodology and the steps conducted to answer the research question and to arrive at the desired classification. These results are elaborated in depth in Sect. 4, which can hence be considered the core of the paper. A discussion of the results, especially with regards to their applicability, is given in Sect. 5, before Sect. 6 summarizes the main conclusions and outlines remaining limitations as well as future research directions.

2   Research Context

2.1   Explorative BPM Techniques

Business Process Management revolves around understanding, managing, and transforming how work is performed in an organization [4]. The foundational understanding of performing work is that it serves as a mechanism to create or add value. BPM specifies this in studying business processes, i.e., end-to-end chains of events and activities that create value through transforming inputs into outputs [11]. For managing business process performance, BPM offers a variety of different methods, techniques and tools [4]. While the BPM discipline has matured significantly both in academia and practice over the past decades, and as such has evolved into a holistic, professionalized management approach [5], the exploitative perspective, i.e., a focus on addressing roadblocks to high performing process performance, has been at its core [1]. Extending the focus to opportunities, so-called explorative BPM, can still be considered a nascent discipline [1]. Though the overall topic of explorative BPM is gaining popularity, as evidenced by dedicated tutorials [6] and even a first proposal for a curriculum (see https://explorative-bpm.com/), the actual body of knowledge on operational and well-defined explorative process design techniques is still rather limited. One comprehensive approach for explorative BPM is the Five Diamond Method, which captures business, innovation, purpose, technologies as well as their overall integration [12]. However, the comprehensiveness of this approach comes with compromises in terms of the specificity of its embedded techniques. Its focus is also on extending the design space of a typical process scope as opposed to the pre-process scope that constitutes process latency.

One of the first, more detailed contributions is the use of explorative patterns [13]. The in total seven patterns generate growth-related process design possibilities for existing business processes from an opportunity-centric lens [13]. These patterns provide deductive guidance on how to expand a business process in the search for new value propositions. However, a restriction to post-action coding of few salient cases and a lack of contextualization limit their validity so far [13]. One pattern, called Process Initiation, suggests the reduction of process latency or 'time-to-process' [13], but it does not provide a sufficient level of detail for operationalization.

2.2 Process Latency

The general concept of latency as an interval between stimulation and response is well known in various disciplines. While there are several definitions of the term, latency commonly denotes the time that passes from the moment an event occurs until an appropriate response is generated and executed [14]. Hence, latency can be characterized as a time period, or more specifically, a reaction time or 'time to initiate'. As such, response latency of probands is used in psychology to examine the strength of certain mental connections and is, e.g., applied in the context of advertising and branding research [15, 16]. In electronics engineering and computer science, response latency refers to the delay of signal processing in a network [17]. Low latency implies that there are no or almost no delays, which is critical for many applications – hence, significant research efforts are being undertaken towards technical latency reduction [18]. One example for low latencies determining successful strategies comes from the world of trading: Low-latency trading activities, where responses to market events are automatically triggered in milliseconds, can create profit opportunities through an advantage of relative speed over other traders [14]. An edge in speed over competitors has been well researched in the business and strategy field as a special case of competitive advantage. As an early contribution, the concept of first-mover advantage formed around timely market entry [19], although its effects need to be interpreted in strong linkage to, among other aspects, the resources possessed by the respective firm and other strategic decisions [20]. Equally emphasizing the value of speed and reduction of lead times to enable, e.g., faster responses, time-based competition strategies summarize internal optimization activities in the entire organization [21–23] and are thus closely related with exploitative BPM approaches in terms of minimizing processing time.

In supporting internal decision making through IT systems and data, the Business Intelligence (BI) discipline adopts a related focus on internal activities. There, the value of timely responses to relevant 'business events' is acknowledged through conceptualizing a so-called 'action distance' [24]. Corresponding to the reaction time for initiating an appropriate action following the event, it may be regarded synonymously to the concept of latency as described above [24, 25]. Furthermore, it is commonly broken down into three components [24]: The data latency occurs after the event has happened and until the data is collected, stored, and ready for analysis. Subsequently, the analysis latency addresses the time it takes for results to be generated and presented from the data. As those two components are mainly driven by the supporting technology, they are summarized as infrastructure latency [25]. Finally, the decision latency denotes the time to initiate a response after the analysis results are available. Driven by the value of timely information and related opportunity costs, as different courses of action may become unavailable the higher the overall latency, BI research has addressed ways to reduce action distance, e.g., through appropriate software systems [9].

Complementary, in the search for low response latencies and fast reaction times, the sensing of signals has a significant role to play. From the introduced definition of latency, it is evident that learning about an event is the key first step. In order to do so, one must be able to identify and interpret the signals these events inevitably send out. Those may range from fundamental body functions, like a raise in heart rate as physical signal corresponding to a certain event, to sophisticated technological signals created by sensors, e.g., from the interruption of a light barrier, which are usually coupled with an underlying interpretation logic. When relying on existing signals, i.e. from events that have already taken place, the key lever lies in optimizing the response: In the case of low-latency trading, the market events and respective signals are commonly known and traders engage in a "technological arms race" to execute appropriate actions the fastest [14]. Same applies for the abovementioned reduction of infrastructure latency, which equally depends on technological factors [24]. Those signals can be considered lagging indicators, referring to them resulting from past actions or events and, thus, being reactive in nature [26].

On the other hand, it becomes increasingly feasible to explore leading indicators that aim to predict future developments and thus, are anticipating that certain events are expected to take place [27]. As there is a natural degree of uncertainty to such anticipations, which becomes greater the further into the future a prediction is to be made [28], leading signals can be considered 'weaker' than lagging ones. When there is not one obvious event, the differentiation of occurring signals in lagging and leading, known for example from performance management [26], is not always well defined, as it requires specifying a certain cutoff moment that serves as the boundary between what would be considered leading or lagging [29]. Nonetheless, moving towards predictive analyses based on sensing of earlier signals has the potential to create new value propositions for both organizations and customers [28]. Organizations may adopt a proactive strategy to conveniently deliver value before customers are even aware of corresponding needs or have expressed them, respectively [30]. While explicit needs still have to be addressed first, for lasting competitive advantage also proactive strategies are needed [30–32]. In such a proactive approach to the customer journey, customers are actively moved and led along processes relevant for them instead of reactively following or reacting to them [33]. For the service delivery model in the context of governmental services, for example, related mechanisms are referred to as "flipping the service delivery model from a reactive pull to a proactive push" [34, 35]. The technological progress in the form of event monitoring and subscription solutions (e.g., in systems such as Salesforce or Oracle, or as demonstrated by Amazon's Dash Replenishment program) makes proactive services attainable and adds to the growing importance of addressing proactivity in the public sector [36].

It is those rapid and continuously evolving technology advancements that drive the pivotal importance of an ability to identify, capture and analyze weak signals from rich datasets and act upon them in virtually all industries [10]. Consequently, access to data and sufficient capabilities for its analysis are indispensable to establishing a 'signal advantage' as a source of competitive advantage in the digital era [8, 10, 28].

Although these aforementioned concepts are generally addressing ways of delivering or adding value and thus, seem inherently relevant to the explorative BPM discipline, an application to process redesign has so far been one-sided on processing time reduction. As such, for business process analytics, event detection and analysis within the boundaries of the execution of a specific process (instance) is fundamental [25]. Meanwhile, the exploration of process latency reduction opportunities, which materialize prior to the process instantiation, has largely been neglected [13]. To provide guidance on related strategies based upon process latency reduction activities, delving into the mechanisms of how successful firms already reap the benefits of related actions seems promising [37].

3 Research Methodology

In order to study ways in which organizations can engage in process latency reduction activities, classifying such cases into different prototypical categories marks an auspicious first step. Such classifications are effective in reducing the complexity of a topic as to enable meaningful analysis [38]. Thus, they are especially useful in disciplines were little knowledge about the subject is yet available [39], as it is the case for explorative BPM and the novel concept of process latency [1, 13]. Consequently, classifications and resulting frameworks are frequently utilized as a tool to describe and structure complex subjects and, as such, have been well researched from a methodological point of view [39].

One key finding is that a framework should be both grounded in theory and informed by empirical observations [38]. While the first aspect should be addressed through carefully studying the extant literature from relevant disciplines, several options exist to include empiricism. Among those, focus groups are a well-defined qualitative research methodology that allows to explore research subjects in detail by having a converging discussion with about 4–12 qualified participants [40]. Hence, the focus group methodology is also referred to as a "group interview technique", by which a substantial amount of rich data can be collected in a comparatively short timeframe [41]. Another key advantage is the openness of the format in combination with the opportunity for the researcher to directly interact with the participants [40]. This is assumed to be vital in a field like explorative BPM, a still nascent discipline, where the degree of novelty is high and thus, the chance of misunderstandings might be, too.

Our applied research methodology involved an iterative approach to developing the final results. The overall step-by-step procedure with three iterations over the course of three months is summarized in Table 1. Each iteration consisted of a series of design activities, performed by the authors, followed by demonstration activities to assess the results of the design phase. Those results take shape as intermediate versions of the aspired classification of process latency reduction strategies. While we refer to the online appendix for the results of iterations one and two, the third and final iteration resulted in the classification presented in the following chapter. To guide the demonstration activities, we relied on criteria from literature for both objective as well as subjective assessment. For example, the categories should be unique with at least one object classified for each of them. In addition, the last iteration before terminating the process should see no dimensions or characteristics merged, split or added. As for subjective criteria, the categories should be constructed in a robust and comprehensive, yet concise way. This combination of attributes is often referred to as MECE, i.e., mutually exclusive and collectively exhaustive. Beyond that, it needs to be explanatory for one to actually capitalize on the findings in subsequent research or application endeavours.

First, we studied the extant literature on the concept of latency and its conceptualization and utilization in various disciplines to develop an initial understanding. The

Table 1. Applied development process

Iteration	1	2	3
Timeline	Jul – Aug 2021	Aug 2021	Aug – Sep 2021
Design activities	• Comprehensive literature study on latency concepts and use • Call for and reception of first set of case examples • Initial analysis and 1st version of classification	• Complementary literature study and identification of add. examples • Revised 2nd version of classification based on new input and discussion from first round of focus groups	• Revised and modified final version of classification as well as generalized information based on discussion from second round of focus groups
Demonstration activities	• First round of focus groups	• Second round of focus groups	• Participant feedback in written form and personal interviews

respective results have been presented in Sect. 2. To move towards a conceptualization of forms of process latency reduction and to enable an understanding of how and which new value is created through them, we then involved a global group of contributors for two main purposes: (1) input of relevant case examples to be classified and (2) participation in focus groups as part of the demonstration activities. The contributors were identified via open invitation among the professional network of the authors on LinkedIn in June 2021. In total, 18 participants expressed an interest to contribute. Each of them was carefully assessed in terms of professional and academic expertise in BPM.

In selecting the overall pool of participants, we were mindful of ensuring geographical and gender balance as well as a good balance between BPM professionals and BPM academics in different stages of their career. In result, 14 participants were invited for contribution of case examples and to take part in the focus group sessions. Table 2 provides an overview about the demographics of the members of our focus groups. As described, a vital part of the design activities within the first iteration was driven by the case examples contributed by the participants. Those served as references for the unit of analysis, the study of reducing process latency. The initial set of case examples was examined and classified by the authors. As a guiding principle, the examples were analyzed on four different levels: (1) What is the activity conducted by the organization in the example, i.e., which action was taken to expand the value proposition, which business process is affected, and which of its elements are added or changed? (2) Why is the activity conducted and how is its success measured? (3) When is a certain activity applicable, i.e., what are related context factors? (4) How do organizations make an activity work, i.e., what are related success factors?

Table 2. Overview of focus group participants

No	Role	Industry	Region	Sex
1	Product Manager	Retail	Middle East	m
2	Industry Analyst	Consulting	North America	f
3	Assistant Professor	Research	Europe	f
4	Managing Director	Consulting	Australia	m
5	Associate Professor	Research	Southeast Asia	f
6	Lead Business Analyst	Banking	Australia	m
7	Head of Digitalization	Chemicals	Europe	m
8	Independent Consultant	Consulting	Australia	m
9	Full Professor	Research	Europe	m
10	Associate Professor	Research	Europe	m
11	Lead Operational Excellence	Chemicals	Europe	f
12	Research Associate	Research	Europe	m
13	Vice President Operations	IT & Software	Australia	f
14	Manager	Consulting	Europe	m

After the completion of the initial design activities, the first evaluation in the form of focus groups was conducted. The sessions were designed and conducted with respect to and based on guidelines from relevant literature [40–43]. As for the second round of focus groups, also this round was broken down into different sessions to cater for time zone preferences and to keep the number of participants within the preferable range. The authors served as moderators for the meetings. The sessions were recorded (participants were asked for written consent anteriorly), transcribed and then analyzed by the authors using an open coding procedure [40, 44], which generated the data basis for the subsequent design activities. Specifically, after the first iteration, the feedback of the participants showed that the categorization was not yet concise and explanatory enough, as became implicitly evident, for example, from several clarifications that were required during the discussions, with participants raising concerns about whether their respective case examples had been classified appropriately.

The design activities in the second iteration comprised of a complementary literature study and a subsequent identification of additional case examples from empirical and theoretical sources, which added to the ones initially contributed by the participants to form the final set of 28 case examples. Informed by the abovementioned feedback from the first round of focus groups, we revised the classification and developed a second version. From that, additional information was elicited through abstraction and generalization, followed by a second round of focus groups which further added feedback and insights.

Consequently, in the third and final iteration, additional modifications to the classification were made and the accompanying generalized information was revised and expanded in the design phase, before the final feedback of potential users, i.e., the focus group participants, concluded the demonstration phase of the third iteration. This materialized in the collection of explicit statements and feedback from the participants on usefulness and usability, two key criteria for developing those kind of categorizations [45]. Feedback was sent in written form, but also issued in two calls with immediate follow-up discussions. The comments and statements led to minor adjustments in wording and additions, however, there were no substantial changes made to the core structure, that will now be presented in the following chapter. This underpinned the decision to end after the third iteration.

4   Process Latency Reduction Strategies

The final outcome represents a detailed description of process latency reduction strategies by means of a set of 28 relevant case examples and their classification into three types. Here, we will only elaborate on these resulting categories, which we refer to as process latency reduction strategies. For more details on the case examples and their classification into the three strategies, please refer to the online appendix.

In general, all cases involve a reduction of process latency, i.e., the time from occurrence of a need to the start of the respective fulfillment process. Hence, the strategies are differentiated based on which individual latency in terms of the generic customer journey they influence. Figure 1 shows a visualization of a schematic customer journey that was used as the starting point for conceptualization, as it is meant to represent the status *before* a process latency reduction activity is conducted.

Fig. 1. Generic customer journey before latency reduction

This understanding of the customer journey in a simplified and abstracted form, which fits the purpose of describing and classifying sub-types, can be found in similar forms in the literature [46]. The understanding of the notion of latency as a time needed to perform such an action is shared with common definitions, as were introduced in Sect. 2. More specifically, as process latency is understood as "time between the occurrence of a demand and the initiation of the related process" [13], the event here can be viewed synonymously to detectable state change implying a demand for action [25]. In our applied definition of the concept, process latency consists of three individual activity latencies. Such decomposition is also employed in the literature as discussed in Sect. 2.2 [24, 25]. Yet, we adopt a significantly different lens on the latency concept by aligning it with the items of the generic customer journey:

$$Process\ Latency = t_{ef} = t_{ea} + t_{ar} + t_{rf} \tag{1}$$

In the abstracted formula, t_{xy} represents an individual latency and hence, denotes the time between x and y. The letters are taken from Fig. 1, with process latency being the time from the event e, i.e., an occurrence of demand, to the start of the related fulfillment process f. As becomes evident from this conceptualization, to reduce process latency means reducing at least one of the individual activity latencies. This insight is central to the classification of process latency reduction strategies presented in Fig. 2. The first type is about the so-called awareness latency t_{ea}, the second one addresses the request latency t_{ar}, while the last one concerns the fulfillment latency t_{rf}.

The awareness latency is reduced when the awareness is moved as close to the start event as possible. For this, the organization needs to detect the start event and proactively approach the customer. The activity *"Create awareness"* is added to the organization's layer of the customer journey. The following parts remain unchanged, as the customer stays entirely in charge of triggering an eventual request. In creating it themselves, the organization can ensure that the awareness happens earlier or even at all.

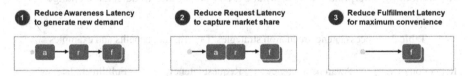

Fig. 2. Process Latency reduction types from a customer journey perspective

By that, new demand is generated, or as one participant expressed it: *"It's about creating new instances that would not have occurred if they hadn't been prompted in some way."* Beyond that, it presents the organization as trusted adviser to the customer, who gains a high degree of convenience. In extreme cases, a short awareness latency can even fulfill preventive functions, especially when a potential unawareness of a need (e.g., for a medical treatment) has severe consequences.

For a reduced request latency, the request is moved as close to the customer's awareness as possible. As the customer detects the start event and is responsible for triggering the request in the end, the role of the organization lies in making the request easy and highly convenient, e.g., by introducing new ways of placing orders. The allocation of activities along the customer journey is not altered. Again, the words of a participant provide a succinct summary: *"This was going to happen. The customer recognized that they needed to do something. It's just reducing the barriers to enable them to do it as quickly as possible."* With a short request latency, the customer is unlikely to reconsider the need or place it with a competitor and will rather choose the 'path of least resistance' instead. For extreme cases, like emergency calls, the request latency is inversely related to the effectiveness of the response. Should, in the context of request latency reduction, a high repetitiveness in customer behavior and increased trust be observed, it might be possible to advance towards the third type.

A reduction of fulfillment latency, i.e., moving the start of the fulfillment process as close to the event as possible, can entail the most significant changes. Ideally, awareness and request evaporate for individual process instances, as the customer has previously given a dedicated form of consent or mandate for a general relationship. It is up to the organization to detect the start event and trigger an appropriate process, which often corresponds to an automated execution. For individual instances, the customer does not need to perform any kind of activity prior to receiving the fulfillment of a conscious or unconscious need. Both organization and customer benefit from the bilateral relative certainty of supply and demand with a maximum level of convenience for the customer. In certain contexts, a process might need to be triggered even when awareness and request are impossible. Here the automated execution can be of especially high value, e.g., when an emergency call can be triggered although the customer itself is unconscious. Table 3 summarizes the introduced types of process latency reduction strategies including their basic descriptions, key benefits along the three categories *Revenue, Customer, New Value* and selected representative examples.

Even within these types, there exist two common sub-types for all of them. These relate to the previously introduced differentiation of lagging and leading signals, which is not universally defined and depends on the choice of the key event that separates the leading from the lagging domain [29]. Much alike, the applied definition of an event leaves room for interpretation as to what exactly constitutes a 'detectable change'.

Table 3. Process latency reduction strategies: Description, benefits and examples

Reduction of...	Awareness latency	Request latency	Fulfillment latency
Description	Awareness is moved as close to event as possible	Request is moved as close to awareness as possible	Fulfillment is moved as close to the event as possible

(continued)

Table 3. (*continued*)

Reduction of...		Awareness latency	Request latency	Fulfillment latency
Benefits	Revenue	Increased demand as new instances are created	Increased market share, as request is taken before competitors might do so	Recurring sales, certainty of demand, higher customer lifetime value
	Customer	Convenience, reduced search costs, builds trust	Significantly easier access to products and services	Extreme convenience, certainty of supply up to prevention of outages
	New Value	Preventive function, if unawareness has major consequences	Immediate request can be critical to successful service delivery	Crucial when awareness and/or request are impossible
Examples		Push notification in case bank detects unusual account activity, context-sensitive proactive suggestions	One-click credit card blocking in case customer detects unusual account activity, voice-enabled ordering	Immediate credit card blocking in case bank detects unusual account activity, continuous replenishment

As a practical solution, the introduced types are classified into sub-types based on whether the respective start event is considered to be the actual occurrence of a need ('immediate awareness/request/fulfillment') or its prediction ('predictive awareness/request/fulfillment'). As previously discussed, such a prediction implies a degree of uncertainty not present in the detection and analysis of past events, making these events and their respective signals 'weaker' in nature.

The examples are distinguished accordingly. To illustrate the approach, consider the first type, i.e., reduction of awareness latency. In the immediate form, the awareness is created right when the potential need occurs. Context-sensitive proactive recommendations, for example when traveling, fall into this category. In the predictive form, a smart wearable device might suggest scheduling an appointment with a medical practitioner based on predictive health data analytics, long before actual symptoms occur. It is evident that especially the latter form requires entirely new signal and event sensing capabilities [10]. Additionally, the earlier a prediction is to be made, the weaker the respective signals will likely be. This raises important questions for the implementation of process latency reduction strategies.

5 Discussion

The presented classification of process latency reduction strategies can help identifying related opportunities that arise from these approaches for an individual organization.

However, it is clear that the applicability of a certain type and thus, its evaluation and implementation need to be regarded in the specific organizational context. Organizations should generally think about whether a process could at all benefit from a lower latency, e.g., because of a non-ideal start event with significant transaction cost, a highly competitive environment with hard-fought market share distribution, or general opportunity cost arising from courses of action that are only available for a certain amount of time [9]. There might very well be processes where a process latency reduction only adds minimal value, if any. At the same time, there will usually be costs associated with a process latency reduction initiative, resulting from, e.g., investments needed for early event sensing or implementation of respective systems. Naturally, a careful evaluation of expected cost and benefit is required to inform a decision on different process latency reduction strategies.

Moreover, to assess initial ideas, especially in situations where the uncertainty is high, qualitative evaluation can be a useful remedy long before quantitative statements are reasonable [45]. It is important to consider relevant context and success factors for this, as numerous pitfalls may exist. Multiple perspectives have to be taken into account [45, 47]. This goes beyond the economic considerations outlined above. Examples include customer acceptance, e.g., with regards to the necessary levels of customer trust that needs to be established, and feasibility consideration, such as the access to relevant data from existing or new sensors, and the related ability to create or detect early and weak signals for a latency reduction to be implemented successfully. Akin to the use of guiding questions in [45], it is not claimed here to present definitive decision guidelines, but means to facilitate users in qualitative reflection.

As such, this work comes with implications for both practice and academia. In practice, organizations eager to engage in opportunity-centric process redesign activities may include the generic types and related examples as a tool for systemically ideating and examining, which of their business processes may benefit from process latency reduction in any form. For academia, the different latency reduction strategies add to the conceptualization of process latency as a novel redesign metric within the still nascent field of explorative BPM.

6 Conclusions, Limitations, and Future Work

Redesigning processes in the future is expected to be significantly more opportunity-driven, as opposed to the problem-centric lens still dominant today. This explorative approach is a response to an environment that is becoming increasingly opportunity- and data rich. However, to turn data into value requires, among other aspects, capabilities to sense and interpret respective signals, as well as to act upon them. A potential source of value from an enhanced sensing ability, that may constitute a signal advantage, lies in the potential to reduce process latency, which we defined as the time from occurrence of a need to the start of the respective fulfillment process. To examine how organizations may systematically be able to identify process latency reduction opportunities, a classification based on extant literature and several real-world case examples was developed, which differentiates three types of process latency reduction strategies depending on which individual activity latency is addressed: By moving the awareness as close to the start event as possible and proactively making customers aware of potential needs, the awareness latency is reduced. For a reduced request latency, the request is moved as close to the customer's awareness as possible by facilitation of request placement. The most significant changes occur when awareness and request evaporate for individual process instances and the automated execution of fulfillment processes reduces the fulfillment latency. For all types, another level of differentiation is introduced by considering reactive or immediate and proactive or predictive strategies.

With process latency representing a novel redesign metric for BPM, naturally, a set of limitations persists that indicate potential future research directions. While the cooperation within an international expert group and a comprehensive view of extant literature made it possible to expand the conceptual and empirical foundation for process latency reduction with regards to [13], there are likely more relevant data to supply an ongoing identification and classification of case examples into the presented strategies. This may also be considered a form of additional evaluation. Furthermore, a rigorous evaluation in a naturalistic setting, e.g., a detailed case study or workshop, could provide insights on whether the prospected use of the classification is indeed helpful in generating opportunity-driven process redesign ideas. To support this, a structured procedural guidance may be developed that embeds the process latency reduction approaches in a holistic methodology. As a next step, also the applicability discussion needs to be elaborated in terms of customer desirability, viability, and feasibility. Advancing the integration with related business model innovation methodologies seems promising for many of the aforementioned directions.

Data Availability. The full list of case examples classified, intermediate versions of the classification within the iterative development approach, as well as a redacted version of the participant feedback to the final result, can be accessed online under https://bit. ly/3GVnheX.

Acknowledgements. The authors sincerely thank all participants of our global focus groups for their time, valuable contributions and the provision of relevant case examples, without which this work would not have been possible.

References

1. Rosemann, M.: Proposals for future BPM research directions. In: Ouyang, C., Jung, J.-Y. (eds.) AP-BPM 2014. LNBIP, vol. 181, pp. 1–15. Springer, Cham (2014). https://doi.org/10. 1007/978-3-319-08222-6_1
2. Nauck, F., Pancaldi, L., Poppensieker, T., et al.: The resilience imperative: Succeeding in uncertain times. McKinsey & Company (2021)
3. O'Reilly, C.A., Tushman, M.: The ambidextrous organization. Harv. Bus. Rev. **82**(74–81), 140 (2004)
4. Dumas, M., et al.: Fundamentals of Business Process Management, 2nd ed. Springer eBook Collection. Springer, Heidelberg (2008). https://doi.org/10.1007/978-3-662-56509-4
5. vom Brocke, J., Mendling, J.: Frameworks for business process management: a taxonomy for business process management cases. In: vom Brocke, J., Mendling, J. (eds.) Business Process Management Cases. MP, pp. 1–17. Springer, Cham (2018). https://doi.org/10.1007/978-3-319-58307-5_1

6. Grisold, T., Gross, S., Röglinger, M., Stelzl, K., vom Brocke, J.: Exploring explorative BPM - setting the ground for future research. In: Hildebrandt, T., van Dongen, B.F., Röglinger, M., Mendling, J. (eds.) BPM 2019. LNCS, vol. 11675, pp. 23–31. Springer, Cham (2019). https://doi.org/10.1007/978-3-030-26619-6_4

7. vom Brocke, J., Baier, M.-S., Schmiedel, T., Stelzl, K., Röglinger, M., Wehking, C.: Context-aware business process management. Bus. Inf. Syst. Eng. **63**(5), 533–550 (2021). https://doi.org/10.1007/s12599-021-00685-0

8. Reeves, M., Lenhard, E., Rodt, M., et al.: Time-Based Competition with Fast Data, Boston Consulting Group (2013)

9. Olsson, L., Janiesch, C.: Real-time Business Intelligence und Action Distance: Ein konzeptionelles Framework zur Auswahl von BI-Software. Wirtschaftsinformatik Proceedings **2015**, 691–705 (2015)

10. Reeves, M., David, S., Deimler, M., et al.: Signal Advantage, Boston Consulting Group (2010)

11. Hammer, M.: What is business process management? In: vom Brocke, J., Rosemann, M. (eds.) Handbook on Business Process Management 1. IHIS, pp. 3–16. Springer, Heidelberg (2015). https://doi.org/10.1007/978-3-642-45100-3_1

12. Grisold, T., et al.: The five diamond method for explorative business process management. Bus. Inf. Syst. Eng. **64**(2), 149–166 (2021). https://doi.org/10.1007/s12599-021-00703-1

13. Akkiraju, R., Sinha, V., Xu, A., Mahmud, J., Gundecha, P., Liu, Z., Liu, X., Schumacher, J.: Characterizing machine learning processes: a maturity framework. In: Fahland, D., Ghidini, C., Becker, J., Dumas, M. (eds.) BPM 2020. LNCS, vol. 12168, pp. 17–31. Springer, Cham (2020). https://doi.org/10.1007/978-3-030-58666-9_2

14. Hasbrouck, J., Saar, G.: Low-latency trading. J. Financ. Mark. **16**, 646–679 (2013). https://doi.org/10.1016/j.finmar.2013.05.003

15. Lowrey, T.M., Englis, B.G., Shavitt, S., et al.: Response latency verification of consumption constellations: implications for advertising strategy. J. Advert. **30**, 29–39 (2001). https://doi.org/10.1080/00913367.2001.10673629

16. Reimann, M., Castaño, R., Zaichkowsky, J., et al.: Novel versus familiar brands: an analysis of neurophysiology, response latency, and choice. Mark. Lett. **23**, 745–759 (2012)

17. Abbosh, O., Downes, L.: 5G's Potential, and Why Businesses Should Start Preparing for It. Harvard Business Review (2019)

18. Schulz, P., Matthe, M., Klessig, H., et al.: Latency critical IoT applications in 5G: perspective on the design of Radio interface and network architecture. IEEE Commun. Mag. **55**, 70–78 (2017). https://doi.org/10.1109/MCOM.2017.1600435CM

19. Lieberman, M.B., Montgomery, D.B.: First-mover advantages. Strat Mgmt J **9**, 41–58 (1988). https://doi.org/10.1002/smj.4250090706

20. Lieberman, M.B., Montgomery, D.B.: First-mover (dis)advantages: retrospective and link with the resource-based view. Strat Mgmt J **19**, 1111–1125 (1998). https://doi.org/10.1002/(SICI)1097-0266(1998120)19:12

21. Stalk, G.: Time—The Next Source of Competitive Advantage. Harvard Business Review (1988)

22. Šapkauskienė, A., Leitonienė, Š: The concept of time-based competition in the context of management theory. Eng. Econ. **21**, 205–213 (2010)

23. Tersine, R.J., Hummingbird, E.A.: Lead-time reduction: the search for competitive advantage. Int. J. Oper. Prod. Manag. **15**, 8–18 (1995). https://doi.org/10.1108/01443579510080382

24. Hackathorn, R.: Minimizing action distance. DM Rev. **12**, 22–23 (2002)

25. zur Muehlen, M., Shapiro, R.: Business process analytics. In: vom Brocke, J., Rosemann, M. (eds.) Handbook on Business Process Management 2, pp. 243–263. Springer, Heidelberg (2015). https://doi.org/10.1007/978-3-642-45103-4

26. Kaplan, R.S., Norton, D.P.: Linking the balanced scorecard to strategy. Calif. Manage. Rev. **39**, 53–79 (1996). https://doi.org/10.2307/41165876

27. Anderson, K., McAdam, R.: A critique of benchmarking and performance measurement. Benchmarking **11**, 465–483 (2004). https://doi.org/10.1108/14635770410557708
28. Davenport, T.H.: A Predictive Analytics Primer. Harvard Business Review (2014)
29. Manuele, F.A.: Leading & lagging indicators: do they add value to the practice of safety? Prof Saf. **12**, 28–33 (2009)
30. Narver, J.C., Slater, S.F., MacLachlan, D.L.: Responsive and proactive market orientation and new-product success. J. Prod. Innov. Manag. **21**, 334–347 (2004). https://doi.org/10.1111/j. 0737-6782.2004.00086.x
31. Blocker, C.P., Flint, D.J., Myers, M.B., et al.: Proactive customer orientation and its role for creating customer value in global markets. J. Acad. Mark. Sci. **39**, 216–233 (2011). https:// doi.org/10.1007/s11747-010-0202-9
32. Sauerwein, E., Bailom, F., Matzler, K., et al.: The kano model: how to delight your customers. In: International Working Seminar on Production Economics, Innsbruck, pp. 313–327, February 1996
33. Edelman, D.C., Singer, M.: Competing on Customer Journeys. Harvard Business Review (2015)
34. Scholta, H., Mertens, W., Kowalkiewicz, M., et al.: From one-stop shop to no-stop shop: an e-government stage model. Gov. Inf. Q. **36**, 11–26 (2019). https://doi.org/10.1016/j.giq.2018. 11.010
35. Linders, D., Wang, C.-M.: Proactive e-governance. In: Janowski, T. (ed.) Proceedings of the 7th International Conference on Theory and Practice of Electronic Governance, pp. 154–157. ACM, New York (2013)
36. Sirendi, R., Taveter, K.: Bringing service design thinking into the public sector to create proactive and user-friendly public services. In: Nah, F.-H.-H., Tan, C.-H. (eds.) HCIBGO 2016. LNCS, vol. 9752, pp. 221–230. Springer, Cham (2016). https://doi.org/10.1007/978-3-319-39399-5_21
37. Brege, H., Kindström, D.: Exploring proactive market strategies. Ind. Mark. Manage. **84**, 75–88 (2020). https://doi.org/10.1016/j.indmarman.2019.05.005
38. Bailey, K.D.: Typologies and Taxonomies: An Introduction to Classification Techniques. Sage, Thousand Oaks, Calif (1994)
39. Gregor, S.: The nature of theory in information systems. MIS Q. **30**, 611–642 (2006). https:// doi.org/10.2307/25148742
40. Tremblay, M.C., Hevner, A., Berndt, D.J.: The use of focus groups in design science research. In: Hevner, A., Chatterjee, S. (eds.) Design Research in Information Systems, pp. 121–143. Springer, US, Boston, MA (2010)
41. Gibson, M., Arnott, D.: The use of focus groups in design science research. In: ACIS 2007 Proceedings 14, pp. 327–337 (2007)
42. Burgess, S.: The use of focus groups in information systems research. Int. J. Interdisciplinary Soc. Sci. Ann. Rev. **5**, 57–68 (2010). https://doi.org/10.18848/1833-1882/cgp/v05i02/51567
43. Belanger, F.: Theorizing in information systems research using focus groups. AJIS **17**, 109–135 (2012). https://doi.org/10.3127/ajis.v17i2.695
44. Williamson, K., Given, L.M., Scifleet, P.: Qualitative data analysis. In: Williamson, K., Johanson, G. (eds.) Research methods: Information, systems and contexts, 2nd edn., pp. 453–476. Chandos Publishing, Oxford (2018)
45. Gilsing, R., Turetken, O., Özkan, B., et al.: Evaluating the design of service-dominant business models: a qualitative method. Pacific Asia J. Assoc. Inf. Syst. **13**, 36–70 (2021)
46. Siggelkow, N., Terwiesch, C.: The Age of Continuous Connection. Harvard Business Review (2019)
47. IDEO: How to Prototype a New Business (2017). https://www.ideou.com/blogs/inspiration/how-to-prototype-a-new-business

Bridging the Gap Between Process Mining Methodologies and Process Mining Practices

Comparing Existing Process Mining Methodologies with Process Mining Practices at Local Governments and Consultancy Firms in the Netherlands

Evelyn Zuidema-Tempel[1,2(✉)], Robin Effing[1], and Jos van Hillegersberg[1,3]

[1] University of Twente, Drienerlolaan5, 7522 NB Enschede, Netherlands
e.tempel@utwente.nl
[2] Saxion University of Applied Sciences, Handelskade 75, 7417 DH Deventer, Netherlands
[3] Jheronimus Academy of Data Science, Sint Janssingel 92, 5211 DA 's-Hertogenbosch, Netherlands

Abstract. This study aims at identifying the differences and similarities between existing process mining methodologies and process mining practitioner experiences. Four existing process mining methodologies are critically reviewed and compared with process mining project elements derived from process mining practitioner experiences and available literature on process mining challenges and enablers. In total 27 interviews with process mining experts of consultancy firms and professionals at local governments have been conducted. Results show that overall existing process mining methodologies lack focus on stakeholder involvement, quantifying and selecting improvement actions, communicating quick wins and results. Also considering organizational commitment and data availability as prerequisites for process mining projects, process selection, vendor- and tool selection, acting on low familiarity with process mining is lacking in various methodologies. Finally, creating a dashboard with flexibility to include self-selected KPIs and metrics, and applying process mining on a continuous basis is considered important by interviewees while is lacking in methodologies. In future research on process mining methodologies it is recommended to take these elements into account. This is expected to give process mining practitioners guidance and support in applying process mining in organizations and stimulate the adoption of process mining in organizations.

Keywords: Process mining · Process mining methodology · Gap analysis · BPM

1 Introduction

Process mining is a technique that is designed to discover, monitor and improve actual processes (i.e. not assumed processes) by extracting knowledge from event logs commonly available in today's information systems [1]. Process mining is used to improve performances of business processes and analyze compliance to business rules [1] and to

C. Di Ciccio et al. (Eds.): BPM 2022, LNBIP 458, pp. 70–86, 2022.
https://doi.org/10.1007/978-3-031-16171-1_5

achieve digital transformation in organizations [2, 3]. This technique is becoming more popular [4] and the process mining market is growing fast. It is estimated that there are approximately 35 process mining vendors offering process mining products and services. The process mining market for new product license and maintenance revenue is valued at $550 million in 2021, a 70% annual growth compared to 2020 [5]. Despite the market growth of process mining, limited research exists on the effectiveness of application of process mining in organizations [4, 6]. The majority of studies on process mining focus on technical aspects of process mining, e.g. developing process mining techniques and improving algorithms [7–9]. There is a clear imbalance between the amount of research conducted on process mining applications on the one hand, and adoption in organizations and the increasing popularity and market growth of process mining on the other hand [4, 10, 11]. Moreover, limited research exists on the application of process mining project methodologies within organizations. These methodologies are important as they give practitioners guidance and support in applying process mining in organizations, stimulate the adoption of process mining in organizations, aid in sharing best practices and prevent reinventing the wheel [12, 13]. Yet, to the best of our knowledge, existing process mining methodologies have only been scarcely validated in one or just a few case studies [12–14]. As a result, it is difficult to assess to what extent existing process mining methodologies effectively support the application of process mining in organizations. Clearly, there is a need for a broader validation of current process mining methodologies.

In this research the following research question has been developed; "What are the differences and similarities between current process mining methodologies and process mining practices at local governments and consultancy firms in the Netherlands?". To answer this research question, four existing process mining methodologies have been critically reviewed and compared to process mining project elements derived from practitioner experiences with process mining. In total, 27 interviews have been conducted with process mining experts of consultancy firms and professionals at local governments. This allows for a broader perspective than the current limited amount of case studies available in literature to validate existing process mining methodologies [12–14]. Based on this comparison, we identify gaps between existing process mining methodologies and practitioners. Recommendations for improvements to (future) process mining methodologies are suggested with the aim to support the enhanced adoption and usage of process mining in practice. The remainder of this paper is structured as follows. In Sect. 2 various process optimization methodologies, process mining methodologies and process mining challenges and enablers are discussed. The used methodology is described in Sect. 3. Section 4 includes the results and Sect. 5 includes the conclusion, limitations and suggestions for future research.

2 Theoretical Background

This section provides an overview of related work on process optimization methodologies, process mining methodologies and enablers and challenges of adoption of process mining in organizations. Given the vast amount of literature on business process optimization and management, we do not strive for a comprehensive overview. Rather, we present a selection of key studies that represent the main lines of research in these areas.

We carried out an extensive search for process mining methodologies and enablers and challenges, resulting in only a limited number of studies available in the literature. This literature survey approach served our aim of illustrating and positioning process mining methodologies and evaluating their use and validation in practice.

2.1 Process Optimization Methodologies

Business Process Reengineering (BPR) can be defined as the fundamental rethinking and radical redesign of business processes to achieve dramatic improvements in critical, contemporary measures of performance such as cost, quality, service and speed [15]. Based on a review on existing BPR methodologies, a BPR methodology for executing BPR in organizations is developed [16]. The methodology analyses the as-is process to identify bottlenecks in the current process, the design of a to-be process, and implementation of the reengineered process. It delivers continuous improvement by initiating ongoing improvement measures. But the method lacks validation by organizations and practitioners. Before starting a process optimization project, organizational readiness and commitment is crucial. A significant need for the process to be reengineered is vital [16]. Also, egalitarian leadership, collaborative working environment, top management commitment, supportive management, and use of information technology are positive indicators when assessing organizational readiness for BPR [17].

Business Process Modelling (BPM) is the discipline that combines approaches for the design, execution, control, measurement, and optimization of business processes. There is a trend to increase focus on process monitoring, adjustment and process diagnosis [1], simplicity, predictions, more extensive cooperation in organizations, anticipating on customer needs and optimizing processes using design-by-doing and optimization iterations [18]. BPM has distinct disadvantages. Factual process data is not always used in redesigning processes and process related decisions. Various stages of BPM are not supported in a systematic and continuous matter and only severe process problems will trigger another iteration in the BPM life cycle when designing or improving the process [1].

Data mining techniques aim to analyze large datasets to find unexpected relationships, and summarize data in an understandable way [19]. The CRoss Industry Standard Process for Data Mining (CRISP-DM) and Sample, Explore, Modify, Model, and Assess (SEMMA) are two widely used data mining methodologies [13]. Both methodologies have limitations regarding the length of the process, selection of data, and needed knowledge on data mining when executing a data mining project [20] and missing guidelines for organizations on how to conduct deployment in data mining projects [21]. SEMMA is considered highly technical and there is a lack of clarity on how to apply new knowledge obtained by data mining. Both methodologies are very high-level, highly complicated [22, 23] and provide little support for process mining methodologies [24]. The focus lies on modelling by using elements such as Petri Nets and analysts are reluctant to use them as they are discouraged by the method complexity, the work needed for preparation of the mathematical model and the difficulty in comprehending and interpreting the results [25, 26].

2.2 Challenges and Enables of Process Mining in Organizations

In order to compare the process mining methodologies identified above with experiences and best practices, we describe process mining challenges and enablers of process mining in organizations. The identified challenges focus mainly on event log and data quality issues [10, 27, 28], selecting appropriate process mining processes [4, 29] and vendors [11], business case calculation and implementation of improvement actions, and making process mining a continuous effort in organizations [4]. These challenges have also been mentioned in the process mining manifesto [30] which also includes improving understandability and usability by non-experts as key challenges. Identified enables of process mining methodologies are actionable insights, confidence in process mining, perceived benefits, and training and development [10], managerial support, project management availability, resource availability, process mining expertise, and data and event log quality [31]. It has to be noted that these process mining challenges and enablers have not been widely empirically validated. Expert views, case studies, surveys or field tests on process mining enablers and challenges are scarce and not systematically studies.

2.3 Process Mining Methodologies

As data mining projects were not tailored towards process mining projects, the L*life cycle model [1] was coined as one of the first process mining methodologies. This model focuses on process mining projects of structured 'lasagna' processes. The Process Mining Project Methodology (PMPM) and the accompanying process mining life-cycle model [12] was designed in response to the lack of process mining methodologies that provide guidance how to apply process mining in practice. The authors of the PM^2 methodology [13] state that the main bottleneck of previous process mining methodologies was the lack of iterative analysis which the authors considered vital. The process mining project proposal [14] was developed as a response to [1, 12, 13]. Previously developed process mining methodological approaches developed provide mostly generic guidelines, but do not define the specific steps and tactics for the challenges that a practitioner must go through when facing a process redesign project through process mining [14]. The L* life-cycle model stages neither reach the necessary level of detail, nor define the specific steps to be followed when it comes to developing a process mining project [14]. The PMPM methodology [12] does not deepen into key aspects such as project planning and data preparation and extraction from the different information systems. The difference of the process mining project proposal compared to previously mentioned methodologies [1, 12] is that it was developed using an engineering design science methodology. Moreover, it was evaluated in three case studies and emphasizes data preparation from different data sources. In Table 1 we present an initial comparison of the methodologies discussed in this section based on their structure, validation, and limitations mentioned in the paper of the respective authors.

Table 1. Initial comparison process mining methodologies.

Methodology	L*Life cycle model of Van der Aalst [1]	PMPM of van der Heijden [12]	PM² of Van Eck, Lu, Leemans, van der Aalst [13]	PM project proposal of Aguirre, Parra and Sepúlveda [14]
Structure	Plan and justify, extract, create control-flow model and connect event log, create integrated process model, operational support	Scoping, data understanding, event log creation, process mining, evaluation, deployment	Planning, extraction, data processing, mining and analysis, evaluation, process improvement/support	Project definition, data preparation, process analysis, process redesign
Validation of use cases	RWS- and WOZ process not specifically linked to the methodology	Invoicing process at Rabobank NL	Purchasing process for spare parts at IBM	Sales/distribution at trading firm, procurement at university, legal advisory process consultancy firm
Limitations	Not mentioned	One use case, limited to invoicing process, understandability, no support choosing process mining techniques	Knowledge transfer. Incorrect filtering and aggregation, not represent actual process. Difficult time-consuming interpretation. Less guidance process selection	No uniformity in the way data sources record business process events. Methodology is perceived technically biased and difficult

3 Methodology

3.1 Practitioner and Expert Interviews

In order to gain in-depth insights in practitioner experiences in process mining projects, semi-structured interviews have been conducted. Semi-structured interviews are suited for gathering independent thoughts, allow for follow-up questions on unclear or interesting answers and aid in examining uncharted territory with unknown possible direction of answer given [32]. As there is limited research and validation conducted on the adoption of process mining and process mining methodologies in organizations [2, 4, 11, 13, 14] semi-structured interviews provide the flexibility to explore this relatively under-researched research topic and aid to the exploratory nature of this research. In total 19 interviews with professionals at 15 local governments are conducted. Using simple random sampling on all Dutch local governments, 9 local governments are selected. The remaining 6 local governments are selected based on the personal network of the researchers. Also, 8 experienced process mining experts working at consultancy firms are interviewed. In total 5 of these interviewees were selected by conducting a Google search on keywords 'Process mining AND experts', and 'process mining AND consultancy'. The remaining 3 interviewees were selected based on the personal network of the researchers.

Using the identified stakeholders of the process mining methodology of [13] and the processes that were analyzed in previous research as identified in Table 1, interviewees were selected based on having knowledge of process steps identified in Table 1, or were analysts responsible for analyzing processes in their organization. As a result, the interviewees have a variety of roles, such as data-analysis, BI analyst, project manager, financial controller, process manager, innovation consultant and process consultant. The interviewees work in organizations of different sizes, ranging from large (500+ FTE), to medium-sized (<500–500 FTE and smalls (<50 FTE). The interviews took place between March 2021 and October 2021. Detailed information on the interviews is available upon request. As only a few studies area conducted into the adoption and usage of process mining [4, 6] and process mining adoption and process mining methodologies can still be considered in its infancy, the interview questions were of explorative nature. Therefore, the interview questions focused on process mining familiarity, desired process insights, process optimization- and mining bottlenecks, involved stakeholders and the steps followed when executing process mining projects. All interviews were transcribed and summarized. Every interview transcript is given an anonymous abbreviation. For the local governments this ranges from LG1 until LG 19, the used abbreviations for the expert interviews are E1 until E8. Parts of the interview transcripts were labeled using the interview topics as described above. Next, similarities and differences were identified between the various labeled interviewee transcripts, also taking into account the role of the interviewees and organizational size.

3.2 Gap Analysis

A gap analysis is a tool or process to identify gaps, or differences between the organization's current situation and expectation, or "what ought to be in place". Gap analysis indicates areas where managers should take action to narrow the gaps between current situation and expectation, hence improving organizational effectiveness [33]. Gap analysis consists of identifying an organization's needs, highlighting the gaps and implement plans to fill the gaps. In this paper, organizational needs are identified by conducting semi-structured interviews, and are deducted from the theoretical framework on current process mining methodologies and its limitations, and process mining enablers and challenges. Using this input relevant elements/criteria of a process mining methodology are identified. These elements/criteria are clustered based on the stages of a process mining project. Using these criteria, a gap analysis has been created in which the four process mining methodologies mentioned in Sect. 2.3 are compared against. Actual implementation of plans lies outside the scope of this paper. The gap analysis aids in systematically assessing the extent to which current methodologies reflect process mining practice in organizations, and which steps or alterations can be made to (future) process mining methodologies to improve connection with process mining practices in organizations.

4 Results

In the following the key outcomes of the semi-structured interviews are gap analysis are included.

4.1 Local Government Professionals

Familiarity with Process Mining
Familiarity and experience with process mining projects is most often seen at large-sized local governments. Process mining projects often initiate out of personal interest in BI- or data analysis, driven by a passion for data and the need and urgency felt to digitalize. Process mining is often at an explorative phase, resulting in isolated efforts only known at BI departments. Process mining pilots often run for years mainly due to data quality issues, lack of trust in process mining and no felt urgency of management in process mining and process optimization. *"It turned out to be very difficult to make a solid business case for process mining, as management only notices the investment, and not the added value and revenue that could be generated. All in all, it took approximately 2 years before we could start" (LG3).* Interviewees working at small-sized local governments often wondered if their organization was ready for process mining. *"The first impressions is that process mining requires a kind of maturity that not all local governments have" (LG8).*

Desired Process Insights and Process Optimization Bottlenecks
The interviewees unanimously mentioned a need for more insights in actual process steps, throughput time and reducing this throughput time to save cost and meet goals in reducing throughput time. Compliance related matters such as segregation of duty and execution of authorized activities were identified as valuable process insights. *"Are people doing the right thing, or are they violating their authorizations? This is important, because we have rules for this in the municipality" (LG9).* Experienced process mining users mostly mined the financial processes because of data availability, expected cost savings and compliance violations, not meeting process KPIs, understandability of the process, and no involvement of many departments. Having knowledge about the process before conducting a process mining project, management support, communicating quick wins and improvement actions, and cooperation between stakeholders is considered crucial and triggers analyses interpretations and improvement actions. The main identified process mining bottlenecks are the time-consuming formatting of the data for process mining, and missing- or incorrect insertion of data in the data source. *"People must be made aware of incorrect data registration and the consequences of not registering the data properly must be communicated to the involved employees" (LG2).* Experienced process miners mention that the tools used only allow for the creation of a single process mining map, while dashboarding with own selected KPIs and continuous

monitoring is preferred. *"We want to monitor improvements in lead time or failure to follow the process in a dashboard, where findings are preferably expressed in time or money savings" (LG12)*. The interviewees mentioned that process optimization is done ad hoc, not at all, based on gut-feeling. Organizational bottlenecks for process mining found are convincing the management board on the added value of process mining, the complex IT landscape with many data sources, the unfamiliarity of process mining in the organization, unclarity about responsibilities, unclarity about the actual process steps due to lack of process step documentation, and finally the lack of organizational urgency to digitalize the processes were mentioned *"It is unknown how these processes are currently monitored, that is hardly done. Not every team is in control. However, we feel pressure to digitize processes and to improve process documentation" (LG7)*.

Stakeholder Involvement and Process Mining Methodologies

Involved stakeholders, or desired to be involved stakeholders mentioned in process optimization- or process mining are the BI analysts who make the process map and conduct process analysis, the management board which approves the investment in process mining resources, the IT department for advising on and implementation of applications in the municipality application landscape and finally the process owners which analyze results and steers improvement actions for their processes Remarkably, besides one local government, the process mining efforts and analysis still remained at the BI department, and a thorough analysis, implementation and setting up improvement actions together with the identified stakeholders did not take place. Process mining methodologies were not specifically followed.

4.2 Process Mining Experts

The financial processes were mostly mined because of understandability of these processes, urgency in the organization to improve the process, Lean Six Sigma projects and the availability of data. *"The starting point is always the company goals, and whether data is available. In reality it turns out that 99% of the times this is the purchasing process" (E2)*. Projects often focused on process discovery to gain insights in actual process steps and throughput time. Reduction of throughput time and compliance-related insights were mentioned, such as authorization of activities and segregation of duty. Enablers of process mining in organizations found are the ability to make a dashboard with own selected KPIs, an affordable purchasing price of tools, simplicity of the tool (no programming) and the ability to execute process mining on a continuous basis. Also mentioned were using correct data, involvement of process owners to being able to interpret the analysis and make changes in the organization, a data-driven culture, support of top-managers, start small to gain trust and support, focus and dedication to the project, link to business goals and communication of results in the organization to create enthusiasm and familiarity with process mining. Using a workshop on (the added value of) process mining to help clients interpret and read a process mining map was considered vital, as often process stakeholders were not familiar with process mining and maps were not considered easy to read and interpret. The bottleneck that was mentioned by all experts was the data quality relating to incorrect data formats and the incomplete insertion of data in systems by employees. *"It was difficult to get the data out of the systems, result-ing in less trust in having good quality available data. The replies were often "we have tried that before but it did not work" (E5)*. Sometimes process mining projects were terminated due to unavailability of data. Other bottlenecks encountered in the execution of a process mining project are the unfamiliarity with process mining hence showing the added value of process mining, identifying process owners, working with fragmented systems and various data sources, and not-well documented process descriptions.

Process mining projects frequently started with an explanation of process mining and its added value to the various stakeholders involved. Research question and KPIs were developed and sometimes linked to business goals. Most time-consuming was the data extraction and cleaning. For the analysis often Disco or ProM was used, but also UiPath and Minit were mentioned. The interviewees favored the ease of use of Disco and the ability to make dashboards with UiPath and Minit, but were less satisfied with some tools' inability to include own selected KPIs in the dashboard, the high purchasing price and the limited perceived ability to execute process mining on a continuous basis. The analysis phase often started with the creation a rough sketch of a first dashboards, followed by designing more detailed dashboards and discussing the outcome with the customer and interpret the analyses. Bottlenecks were listed, and it was determined together with the process owner which improvement actions would yield most result (in time- and cost saving) and took least time to implement. After selection the improvement actions, the improvements were implemented and monitored. At various stages of the project, analysts, process owners and management was involved. What stands out is that several iterative customer validation rounds took place at various project stages. Validation was related to whether customers could relate to the data, to verify if the dashboard answered the customer question and was understandable, and regarding the conclusions drawn from process mining analyses. Validation took place multiple times and has a iterative nature *"We do not know the process as well as the process owners. At the beginning we draw conclusions that were not recognized by the process owners. Together with the process owner we validate whether they recognize our findings. This is an ongoing iterative process" (E8)*.

4.3 Elements of Process Mining Methodologies

Combining the findings in the literature reported on in Sect. 2 and the interviews, the elements that can be considered relevant in process mining projects can be derived (see Table 2). The elements in *italic* were only identified in the interviews, the remaining elements were identified in both the interviews and the theoretical section of this paper.

Table 2. Relevant elements in a process mining project

Cluster	Element
Before starting a process mining analysis	Organizational willingness, *Data availability, Stakeholder involvement*, Linking business goals to PM projects, Process vendor selection, Process selection, Process mining project goal, Desired insights and *KPI selection*, Familiarity with process mining
Process mining analysis	Data extraction, Data preparation, Creation of *dashboard,* Analysis of *dashboard,* Interpretation of *dashboard*, Drawing conclusions
Improving processes	Defining- *Quantifying-* Selecting and Monitoring improvement actions
Project aspects	*Communicating business value using quick wins and results,* continuous effort, Iterative nature, validation

4.4 Comparison Table Gap Analysis

To identify differences and similarities between existing process mining methodologies and process mining practitioner experiences, the identified elements of Table 2 are used to compare the four identified process mining methodologies of the theoretical framework on. Table 3 includes an overview on the differences and similarities of the identified process mining methodology elements against the existing process mining methodologies.

4.5 Gap Analysis

All methodologies include determining a project goal, formulating a problem definition and research questions, and data extraction and preparation to be suitable for process mining. Analyzing results and defining improvement actions is present in all methodologies. Not all methodologies focus explicitly on the involvement of stakeholders in process mining projects and defining improvement actions. [1] focuses mainly on stakeholder involvement in the phase where the project goals and questions are derived and [12] includes theoretical scenarios of stakeholder involvement and mentions involving an analysis, project leader and manager in the evaluation phase. Involvement of analysts, process owners, managers and IT specialists is more included in [13, 14]. Not all methodologies focus on KPI specification and interpretation and drawing conclusions. Only [14] strongly focus on the quantification and selection of improvement actions. Findings in the theoretical framework [16, 17] and the interviews indicate that organizational willingness and data availability are crucial. Lack of these elements can even lead to the termination or not setting up of process mining projects. It therefore is questionable if and to what extent every organization is suitable for process mining projects. Also, hardly any attention is paid to process vendor- tool- and process selection while these

Table 3. Comparison framework process mining methodologies

Element	L*Life-cycle model [1]	PMPM [12]	PM^2 [13]	PM project proposal [14]
Org. willingness	Not specifically addressed	Not specifically addressed	Not specifically addressed	Not specifically addressed
Data availability	Not specifically mentioned	Mentioned in data-understanding, data needs to be available	Purchasing process selected because of good data quality	Not specifically addressed
Stakeholder involvement	After initiating the project, event data, objectives and questions need to be extracted from systems, domain experts and management	Theoretical scenarios in various sectors with doctor, dept. manager, project team, data specialist, employee. Roles not specifically included insteps in methodology. Roles of process miner, project leader, process manager in step 5 in evaluation step. Role of project initiator mentioned in case study	In stage 1: Planning the activity composing team: business owner, business experts, system expert (IT) and process analysis. Business expert and process expert are most important and part of step 1, 3 and 5. Analysis done by analyst	Stakeholder insight via process scope diagram, project definition and data localization based on interviews stakeholders. Data preparation performed with personnel of company to understand process flow and localize data. Improvement actions defined with personnel of the company
Linking business goals to PM projects	Goals/questions to be extracted from systems, domain experts, management	Question-driven projects (link KPIs to process mining project)	Identifying research questions but not linked to business goals	Only mentions that PM projects must impact business performance indicators
Vendor selection	Not specifically addressed	Step A3, no criteria for vendor selection.Focus on Disco/ProM	PRoM used in the case study	Not specifically addressed

(continued)

Table 3. (*continued*)

Element	L*Life-cycle model [1]	PMPM [12]	PM² [13]	PM project proposal [14]
Process selection	Not specifically addressed	Process identification mentioned but refers to understanding the process and which part of the process is logged	Activity in step 1 and achievability of results is influenced by process characteristics and quality of event data	Not specifically addressed
Project goal	3 types of projects: data-driven, question-driven and goal-driven	Step A2l determine the objective	Determining research questions	Focus on problem definition of process and definition of objective/questions to be solved
Desired insights and KPI selection	Stage 4 mentions detect, predict and recommend as activities	Stage A2 objective determination based on discovery, conformance, enhancement	No KPIs, focus in stage 4 on discovery, conformance, enhancement analytics	Not specifically addressed
Familiarity with process mining	Start with question-driven projects when organizations do not have experience	Not mentioned	IBM case; a basic understanding of PM is beneficial for all involved in evaluation	Not mentioned
Data extraction and preparation	Stage 1, process of getting raw data into suitable event logs described. Prep. described	Event log creation. Select data in terms of context, time frame, aspects. Challenges on amount of data and tool available. Prep. in cleaning, constructing, merging and formatting	Stage 2 extraction, stage 3 preparation	Part of data prep. stage, from the source system. Data extraction to a csv file. Preparation at stage 2 data-localization, extraction, quality analysis, cleaning, data transformation

(continued)

Table 3. (*continued*)

Element	L*Life-cycle model [1]	PMPM [12]	PM^2 [13]	PM project proposal [14]
Creation of process dashboard	Not specifically addressed	Not specifically mentioned	Not specifically mentioned, in stage 4 process analytics is mentioned	Not specifically addressed
Analysis of dashboard	Activities of stage 2 (extract)	Soundness, validation in terms of fitness, precision, generalization and structure, Accreditation by initiator of the project to evaluate whether results are interesting for business goals	Stage 4 with 4 activities done by process analyst	Process discovery (actual steps + execution) performance analysis on cycle time and rework and bottlenecks, and social network analysis (relationship between resources and activities)
Interpretation and conclusion	Diagnose after stage 2, conclusion mentioned after stage 3 and 4 as redesign and adjust and recommend	Accreditation step	Diagnose and focus on understanding the discovered process model, conclusion not specifically mentioned	Not specifically addressed
Defining improvement actions	Mentioned	Stage 5 identification on how process can be improved by improvement actions using improvement actions	Process modifications is separate project and different area of expertise. Improvements measured in another project	Mentioned in stage 4L identifying and prioritizing actions
Quantify, select, monitor improvements	Not specifically addressed	Monitoring addressed in step A6	Not specifically addressed	Prioritizing improvement alternatives is mentioned

(*continued*)

Table 3. (*continued*)

Element	L*Life-cycle model [1]	PMPM [12]	PM² [13]	PM project proposal [14]
Communicating quick wins/results	Not specifically addressed	Last step of the process is presenting the project results to the organization	Not specifically mentioned, but can be derived from verification and validation phase	Not specifically addressed
Continuous effort	Not specifically addressed	A16: decide on an elaboration of the process mining project	Improvement expected to occur in specific improvement project	Not specifically addressed
Iterative nature	After stage 2, 3, 4 new or adjusted KPIs of objectives can emerge	alter scope based on data understanding, event log after process mining, conduct analysis after evaluation	Iterative nature of refining research questions, data processing, mining & analysis and evaluation	No iterative steps, methodology follows a linear nature
Validation	Not specifically addressed	Verification of data in system's log on trustworthy, completeness, semantics, safeness. Verification of model with map, validating on representing the real process and accreditation	Verify findings to original data and system implementation. Validate findings to claims of stakeholders. Identify root causes and design ideas for improvement	No specific validation or verification steps mentioned. Working with people who perform the process to ensure data corresponds to the actual execution of the process

are identified process mining challenges in the theoretical framework [4, 11] and in the interview results. Often Disco or ProM was used as a tool, but the choice of vendor and tool selection lacks argumentation. And this is remarkable, as there are approximately 35 process mining vendors with associated tools [5]. Establishing familiarity with process mining and efforts to increase the familiarity of process mining was only mentioned at [13], while the theoretical framework [10, 30, 31] and the interview findings indicate the challenge regarding the low level of familiarity with process mining in organizations. Finally, the results indicated that creating a process mining dashboard with the flexibility to include self-selected chosen KPIs and process mining projects as a continuous effort in the organization important is important, but is not part of the process mining methodologies. The methodologies of [1, 12] share least similarities with the process mining project elements. These methodologies are relatively theoretical, do not explicitly identify various stakeholders in various stages of process mining projects, do not focus much on vender- and process selection, quantification and selection of improvement actions and especially [1] focus least on iteration, validation and process mining as a continuous effort. Because of the iterative nature, stakeholder involvement, various validation efforts and focus on quantification and selection of improvement actions, the methodologies of [13, 14] have most similarities with process mining practitioner's experience. But none of the methodologies include all relevant elements as identified in this research.

5 Conclusion and Discussion

This study identifies the differences and similarities between existing process mining methodologies and process mining practitioner experiences. Similar elements identified at both the practitioner experiences and the methodologies are goal determination, problem definition and research questions, data extraction and preparation, analyzing results and defining improvement actions. However, none of the existing process mining methodologies include all process mining project elements as identified during this research. These are among others tool-, vendor- and process selection, organizational willingness, communication of quick wins, and quantification and selection of improvement actions. This research contributes to understanding the gap between process mining methodologies and practitioner experiences. The first limitation of this study concerns the generalizability. The interviews were held with professionals and process mining experts at local government agencies and consultancy firms. Hence, we provided only a partial view on process mining initiatives. More research on process mining experiences is needed adding to the completeness of process mining experiences and relevancy process mining methodologies. In addition, general recommendations were made to process mining methodologies indicating that there is one process mining methodology, while different type of organizations might benefit from different process mining methodologies. Therefore we recommend the in the future to be developed process mining methodologies to be validated with more case studies in various sectors in order to increase the generalizability of process mining project elements and methodologies identified in this research. It is expected that including these elements derived from practitioner experiences will aid in giving practitioners guidance and support in applying process mining in organizations, stimulate the adoption of process mining in organizations, provide support in overcoming currently identified process mining challenges and prevent reinventing the wheel.

Acknowledgments. We would like to thank the interviewees for sharing their process mining experiences with us. We also would like to thank the Twente Regio Deal, Saxion University of Applied Sciences and Infotopics for support and funding.

References

1. Van der Aalst, W.M.P.: Process Mining Data Science in Action. Springer, Heidelberg (2011). https://doi.org/10.1007/978-3-662-49851-4
2. Boenner, A.: Bayer: process mining supports digital transformation in internal audit. In: Reinkemeyer, L. (ed.) Process Mining in Action. Springer, Cham (2020). https://doi.org/10.1007/978-3-030-40172-6_19
3. Martins, G.: Association for information systems AIS electronic library (AISeL) process mining and digital transformation of organizations: a literature review (A mineração de processos e a transformação digital nasorganizações : uma revisão da literatura Process m) (2020)
4. Grisold, T., Mendling, J., Otto, M., vom Brocke, J.: Adoption, use and management of process mining in practice. Bus. Process Manag. J. **27**, 369–387 (2021)
5. Kerremans, M.: Market guide for process mining. Gartner, 1–33 (2020)
6. vom Brocke, J., Jans, M., Mendling, J., Reijers, H.A.: Call for papers, issue 5/2021: process mining at the enterprise level. Bus. Inf. Syst. Eng. (2020)
7. Augusto, A., et al.: Automated discovery of process models from event logs: review and benchmark. IEEE Trans. Knowl. Data Eng. **31**, 686–705 (2019)
8. Yeshchenko, A., Di Ciccio, C., Mendling, J., Polyvyanyy, A.: Comprehensive process drift detection with visual analytics. In: Laender, A.H.F., Pernici, B., Lim, E.-P., de Oliveira, J.P.M. (eds.) ER 2019. LNCS, vol. 11788, pp. 119–135. Springer, Cham (2019). https://doi.org/10.1007/978-3-030-33223-5_11
9. Dann, D., Teubner, T., Wattal, S.: Call for papers, issue 5/2022. Bus. Inf. Syst. Eng. **63**(2), 213–214 (2021). https://doi.org/10.1007/s12599-021-00687-y
10. Syed, R., Leemans, S.J.J., Eden, R., Buijs, J.A.C.M.: Process mining adoption. In: Fahland, D., Ghidini, C., Becker, J., Dumas, M. (eds.) BPM 2020. LNBIP, vol. 392, pp. 229–245. Springer, Cham (2020). https://doi.org/10.1007/978-3-030-58638-6_14
11. Turner, C.J., Tiwari, A., Olaiya, R., Xu, Y.: Process mining: from theory to practice. Bus. Process Manag. J. **18**, 493–512 (2012)
12. Van der Heijden, T.H.C.: Process mining project methodology: developing a general approach to apply process mining in practice, p. 85 (2012)
13. van Eck, M.L., Lu, X., Leemans, S.J.J., van der Aalst, W.M.P.: PM^2: a process mining project methodology. In: Zdravkovic, J., Kirikova, M., Johannesson, P. (eds.) CAiSE 2015. LNCS, vol. 9097, pp. 297–313. Springer, Cham (2015). https://doi.org/10.1007/978-3-319-19069-3_19
14. Aguirre, S., Parra, C., Sepúlveda, M.: Methodological proposal for process mining projects. Int. J. Bus. Process Integr. Manag. **8**, 102–113 (2017)
15. Hammer, M.: Reengineering the Corporation a Manifesto for Business Revolution. Harper Collins, New York (1993)
16. Muthu, S., Whitman, L., Cheraghi, S.H.: Business process reengineering: a consolidated methodology. Manuf. Eng. (1999)
17. Abdolvand, N., Albadvi, A., Ferdowsi, Z.: Assessing readiness for business process reengineering. Bus. Process Manag. J. **14**, 497–511 (2008)
18. Paschek, D., Ivascu, L., Draghici, A.: Knowledge management – the foundation for a successful business process management. Procedia - Soc. Behav. Sci. **238**, 182–191 (2018)
19. Hand, D., Mannila, H., Smyth, P.: Principles of Data Mining. A Comprehensive, Highly Technical Look Math Science Behind Extract Useful Information from Large Databases, p. 546 (2001)
20. Selamat, S.A., Prakoonwit, S., Sahandi, R., Khan, W., Ramachandran, M.: Big data analytics—a review of data-mining models for small and medium enterprises in the transportation sector. Wiley Interdiscip. Rev. Data Min. Knowl. Discov. **8**, e1238 (2018)

21. Schröer, C., Kruse, F., Gómez, J.M.: A systematic literature review on applying CRISP-DM process model. Procedia Comput. Sci. **181**, 526–534 (2021)
22. Tsakalidis, G., Vergidis, K.: Towards a comprehensive business process optimization framework. In: Proceedings - 2017 IEEE 19th Conference on Business Informatics, CBI 2017, vol. 1, pp. 129–134 (2017)
23. Zhou, Y., Chen, Y.: Project-oriented business process performance optimization. In: SMC 2003 Conference on Proceedings. IEEE International Conference on Systems, Man and Cybernetics. Conference Theme - System Security and Assurance (Cat. No. 03CH37483), vol. 5, pp. 4079–4084 (2003)
24. Van der Aalst, W.M.P.: Process Mining Data Science in Action. Springer, Heidelberg (2016). https://doi.org/10.1007/978-3-662-49851-4
25. Vergidis, K., Turner, C.J., Tiwari, A.: Business process perspectives: theoretical developments vs. real-world practice. Int. J. Prod. Econ. **114**, 91–104 (2008)
26. Ahmadikatouli, A., Aboutalebi, M.: New evolutionary approach to business process model optimization. In: IMECS 2011 - International MultiConference Engineering and Computer Science 2011, vol. 2, pp. 1119–1122 (2011)
27. De Weerdt, J., Schupp, A., Vanderloock, A., Baesens, B.: Process mining for the multi-faceted analysis of business processes - a case study in a financial services organization. Comput. Ind. **64**, 57–67 (2013)
28. R'Bigui, H., Cho, C.: The state-of-the-art of business process mining challenges. Int. J. Bus. Process Integr. Manag. **8**, 285–303 (2017)
29. Thiede, M., Fuerstenau, D., BezerraBarquet, A.P.: How is process mining technology used by organizations? A systematic literature review of empirical studies. Bus. Process Manag. J. **24**, 900–922 (2018)
30. van der Aalst, W., et al.: Process mining manifest. In: Daniel, F., Barkaoui, K., Dustdar, S. (eds.) BPM 2011. LNBIP, vol. 99, pp. 169–194. Springer, Heidelberg (2012). https://doi.org/10.1007/978-3-642-28108-2_19
31. Mans, R., Reijers, H., Berends, H., Bandara, W., Prince, R.: Business process mining success. In: ECIS 2013 – Proceedings of 21st European Conference on Information Systems (2013)
32. Newcomer, K.E., Hatry, H.P., Wholey, J.S.: Handbook of Practical Program Evaluation. Wiley, Hoboken (2015)
33. Peltier, T.R.: Gap analysis. Inf. Secur. Risk Anal. 116–127 (2021)

Process Mining

Systems Mining with HERAKLIT: The Next Step

Peter Fettke[1,2](✉) and Wolfgang Reisig[3]

[1] German Research Center for Artificial Intelligence (DFKI), Saarbrücken, Germany
`peter.fettke@dfki.de`
[2] Saarland University, Saarbrücken, Germany
[3] Humboldt-Universität zu Berlin, Berlin, Germany
`reisig@informatik.hu-berlin.de`

Abstract. We suggest *systems mining* as the next step after process mining. Systems mining starts with a more careful investigation of runs, and constructs a detailed model of behavior, more subtle than classical process mining. The resulting model is enriched with information about data. From this model, a system model can be deduced in a systematic way.

Keywords: Systems composition · Data modeling · Behavior modeling · Composition calculus · Algebraic specification · Systems mining

1 Introduction

Classical process mining methods as established in theory and practice start out with *event logs*, generated by *processes* during their dynamic progression [1,7]. Process mining is designed first of all to discover processes by extracting knowledge from event logs. Each event in an event log is conceived as an activity that has been performed in the process at the point in time given in the event log, and is related to a particular case. Typically, the events of a case are totally or weakly ordered and can be seen as an execution or run of the process.

The left side of Fig. 1 depicts the standard formal approach for understanding an event log, a processes model, and a system, namely, behavior can be understood as three different sets of symbol sequences [3]. In this paper, we propose to follow a different route: there is no reason to assume that the events of a run are totally or weakly ordered. Of course, a clock outside the run may timestamp a run's events. This induces an order; however, this order is irrelevant for a proper understanding of a run. To the contrary, it spoils the *causal* order of events, which orders two events a and b by $a < b$ if and only if a is a prerequisite for b. Of course, $a < b$ implies each potential clock to timestamp a before b. But a timestamped before b only implies that b is not a prerequisite for a. Or, in one sentence: causality matters!

Additionally, systems to be mined are typically not monolithic, amorphous or unstructured, but can best be described and understood as the composition

C. Di Ciccio et al. (Eds.): BPM 2022, LNBIP 458, pp. 89–104, 2022.
https://doi.org/10.1007/978-3-031-16171-1_6

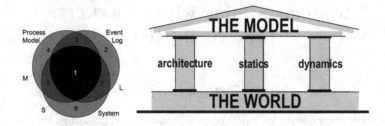

Fig. 1. Classical process mining (left, source: [3]) and systems mining.

of different sub-systems. Hence, an understanding of the different modules of a system is necessary while understanding the behavior of the system. Again, in one sentence: composition matters!

Last, but not least, a system processes data. Data processing is not only needed for the correct execution of processes, but also for the symbolic representation of important objects, e.g. invoices, customers, agreements, orders, products, and many more objects of interest. These objects need to be understood and represented adequately while mining a system. Again, in one sentence: objects matter!

In this contribution, by means of an example, we show how to mine not only process models, but entire system models. This includes the integrated modeling of architecture, statics, and dynamics of the world we live in (Fig. 1, right). To this end, we combine the well-known techniques of Petri nets and abstract data types with the recent *composition calculus*.

We motivate and exemplify a different notion of *runs*, by means of a case study from the area of retail sale. Supported by some static aspects of a system, such runs can be deduced from the system's event logs. Note, that our paper is purely conceptual. We do not provide an algorithm nor a software tool for systems mining. Instead, our main contribution is the elaboration of the new idea of systems mining based on the formal framework of HERAKLIT [5,6].

This paper starts with the presentation of the main idea of modules and runs while unfolding a running case study (Sect. 2). Sect. 3 describes systems nets and Sect. 4 presents the main idea for mining a system module. Related work is discussed in Sect. 5, Sect. 6 presents some conclusions.

2   Modules and Their Composition

Before presenting the (not too heavy) formal framework, we discuss a motivating example that later will be extended to a full case study.

2.1   Example: Occurrence Modules of a Retail Business

We start with a small log, recording observations from the field, namely seven events from a retail shop, as Fig. 2 shows. Each event has a unique name, a set of

Name of the event	involved agents	involved data	time stamp	Name of the event	involved agents	involved data	time stamp	Name of the event	involved agents	involved data	time stamp
shirt to take home	vendor V1 Alice	shirt	1	Alice pays	cashier Alice	50€ receipt	3	Bob pays	cashier Bob	170€ voucher	4
shoes to be ordered	vendor V2 Bob	shoes	2	hat not on offer	vendor V2 Claire	-	4	handing over	vendor V1 Alice	receipt packed shirt	5
V1 packs shirt	vendor V1	shirt	1								

Fig. 2. Event log with seven events.

involved agents, a set of data, and a timestamp. Static inspection of the system and the events of the log identifies *six agents*: Two vendors $V1$ and $V2$, a cashier, and three clients, Alice, Bob, and Claire. All events of the log, up to $V1$ *packs shirt*, include two agents. For example, the *shirt to take home* event includes the vendor $V1$ and the client Alice, jointly selecting a shirt for Alice. The *shoes to be ordered* event includes the vendor $V2$ and the client Bob, jointly agreeing on shoes, to be ordered from wholesale. The other events are intuitively obvious.

From the perspective of agents it is intuitively obvious that for a given event log, an agent is involved in a *sequence* of events, describing one of the potential *behaviors* of the agent. For example, the event log in Fig. 2 implies the vendor $V1$ be involved in three events: *shirt to take home*, *V1 packs shirt*, and *handing over*. This behavior can automatically be deduced from the log. An event updates the local state prior to its occurrence, and produces a local state as a result of its occurrence. Technically, we represent this as a Petri net, as in Fig. 3 (a). Each place (circle) denotes a local state; each transition (rectangle) denotes a step. An agent's behavior deduced from a log is very simple in structure; it can be thought of as a classical sequence of states and steps.

Figure 3 shows the behaviors of all six agents, deduced from the log in Fig. 2. Obviously, they are tightly interrelated, and this interrelation is now to be constructed explicitly. To do this, each behavior is embedded into a *module* in which each transition is either inside the module, or in an *interface* of the module. Each module has a *left* and a *right* interface. Graphically, a module is enclosed in a rectangle with the left and right interface elements on the left and right margin, respectively. The left and right interface of a module A is designated $*A$ and A^*, respectively. This way, Fig. 4 (a) shows the module $V2$ of the vendor $V2$. The two transitions of the module are both located on the right interface, $V2^*$. The *Claire* module in Fig. 4 (b) places the transition of Claire's module on the left interface, $*Claire$.

The two modules are now composed into a new module, $V2 \bullet Claire$, shown in Fig. 4 (c). To compose $V2$ and *Claire*, we merge the transition with label *hat not on offer* of $V2^*$ with the equally labeled transition of $*Claire$. The resulting transition goes inside $V2 \bullet Claire$. The transition with label *shoes to be ordered* of $V2^*$ goes to $(V2 \bullet Claire)^*$.

This example shows the general principle of the composition of two modules A and B: Equally labeled elements of A^* and $*B$ are merged and go into the interior of $A \bullet B$. The other elements from A^* and $*B$ go to $(A \bullet B)^*$ and $*(A \bullet B)$,

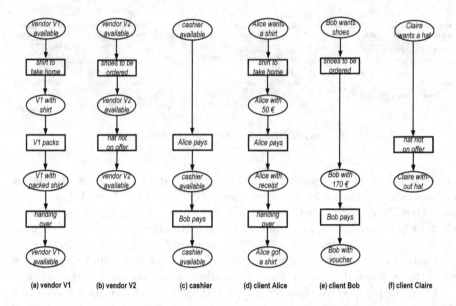

Fig. 3. Agents and their behavior, elicited from the event log.

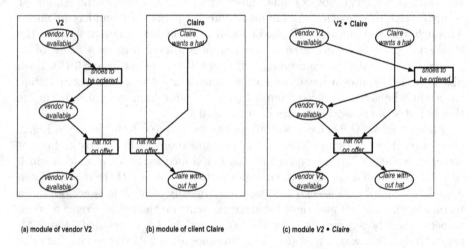

Fig. 4. Two modules and their composition.

respectively. This kind of composition motivates the distinction of right and left interfaces: The running example exhibits an intuitive dichotomy between shop modules (vendors and cashiers), and client modules. Shop modules interact with client modules, so the interface elements of shop modules and of client modules complement each other.

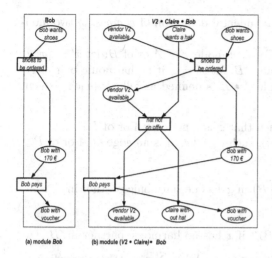

Fig. 5. Composing another module.   **Fig. 6.** Claire • Bob.

To continue, Fig. 5 (a) shows the module *Bob* of the client Bob from Fig. 3 (e). As with the *Claire* module, its transitions lie in its left interface. We now compose $V2 \bullet Claire$ with *Bob* and obtain the module $(V2 \bullet Claire) \bullet Bob$ in Fig. 5(b). Alternatively, we could have formed the module *Claire•Bob* first (Fig. 6) and then module $V2 \bullet (Claire \bullet Bob)$. It is easy to see that the modules $(V2 \bullet Claire) \bullet Bob$ and $V2 \bullet (Claire \bullet Bob)$ are identical. We will see that in general, the composition operator $\bullet$ is *associative*.

2.2 The Formal Framework of Modules

As usual, we represent a Petri net as a triple $(P, T; F)$. We employ the usual graphical representation with boxes, circles, and arrows. In this section, we recall a special case of the composition calculus, and particularly occurrence modules and their composition. The general case can be found in [5].

An *interface over* a set Λ *of labels* is a finite set R, with each element of R carrying a label of Λ. We refrain from the general case of two or more equally labeled interface elements here.

For two interfaces R and S, equally labeled elements $r \in R$ and $s \in S$, are a *harmonic pair* of R and S. A harmonic pair is labeled by the label of r and s. The element s is a *harmonic partner* of r in S, and r is a *harmonic partner* of s in R.

A *module* is a Petri net $N = (P, T; F)$ together with two interfaces $*N$ and $N^* \subseteq P \cup T$, denoted as the *left* and the *right interface* of N. Nodes not in an interface belong to the *interior of* N.

In graphical representations, the interior of N is surrounded by a box, with the elements of the left and the right interface on its left and the right margin, respectively, e.g. Fig. 4.

We are now prepared for the fundamental definition of *composing* two modules:

Let A and B be two modules. For each node x of A or of B, let $x' = \{x, y\}$ if $\{x, y\}$ is a harmonic pair of A^* and $^*B$; let $x' = x$ if no harmonic pair of A^* and $^*B$ contains x. Then the module $A \bullet B$ is defined as follows (each element retains its label):

1. The nodes of $A \bullet B$ are all x' such that x is a node of A or of B.
2. The edges of $A \bullet B$ are all (x', z'), such that (x, z) is an edge of A or of B.
3. The left interface $^*(A \bullet B)$:
   (a) $^*A \subseteq {}^*(A \bullet B)$;
   (b) For $x \in {}^*B$ holds: $x \in {}^*(A \bullet B)$, if x has no harmonic partner in A^*.
4. The right interface $(A \bullet B)^*$:
   (a) $B^* \subseteq (A \bullet B)^*$;
   (b) For $x \in A^*$ holds: $x \in (A \bullet B)^*$, if x has no harmonic partner in $^*B$.

Figures 4, 5 etc. show compositions of modules. Notice that, according to this definition, $^*(A \bullet B)$ or $(A \bullet B)^*$ may acquire different elements with equal labels. However, this never happens in this paper's examples; further details can be found in [5].

A fundamental property of composition is *associativity*, decisive for the usability of modules and their composition. In fact, for any three modules A, B and C holds:

$$(A \bullet B) \bullet C = A \bullet (B \bullet C). \tag{1}$$

As a consequence, it makes sense to just write $A \bullet B \bullet C$. This property is a special case of a more general notion of modules and their associative composition, as discussed in [11].

Furthermore, there are clear criteria for the case of commutativity: for modules A and B holds

$$A \bullet B = B \bullet A \tag{2}$$

if and only if no label occurs in $^*A \cup A^*$ as well as in $^*B \cup B^*$.

The nets in the examples of Sect. 2.1 all exhibit a particular structure: The arcs form no cycles, and each place has at most one ingoing and one outgoing arc:

A net $N = (P, T; F)$ is an *occurrence net* if and only if:

1. The transitive closure of F, usually written as F^+, is a strict partial order, viz. irreflexive and transitive, on $P \cup T$. We denote this relation as $<_N$;
2. for each $p \in P$ there exists at most one arc shaped (t, p) and at most one arc shaped (p, t).

A module is an *occurrence module* if and only if the underlying net is an occurrence net.

For two occurrence modules A and B, the composed module $A \bullet B$ is in general not an occurrence module again. Figure 7 shows an example. This example shows that the interior of A and B matters for this problem. Nevertheless, it can be reduced to a problem of the induced order of interface elements:

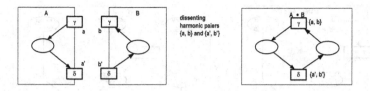

Fig. 7. Dissenting pairs.

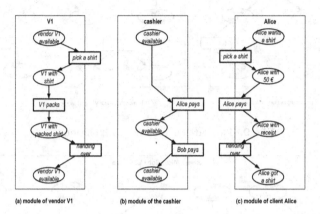

Fig. 8. Three further modules.

With $a, a' \in A$ and $b, b' \in B$, let $\{a, b\}$ and $\{a', b'\}$ be harmonic pairs of A^* and $^*B$. They *dissent* if and only if either $a <_A a'$ and $b' <_B b$, or $a' <_A a$ and $b <_B b'$.

Then, for two occurrence modules A and B it holds: $A \bullet B$ is an occurrence module if and only if A^* and $^*B$ have no dissenting harmonic pairs. All compositions of occurrence modules in this paper yields an occurrence module again.

2.3 Completing the Example

We extend the example of Sect. 2.1 by modules for the remaining three agents of Fig. 3, as in Fig. 8. Figure 9 shows compositions of these modules. Interestingly, the module *cashier* $\bullet$ *Alice* in Fig. 9(a) is an example of a module with elements in both the left and right interfaces. Finally, the module in Fig. 9(b) composes all three modules.

We can now compose the composed module in Fig. 9(b) with the composed module in Fig. 5(b), and obtain the composed module

$$V := V1 \bullet cashier \bullet Alice \bullet Bob \bullet V2 \bullet Claire \tag{3}$$

in Fig. 10. The two interfaces of this module do not contain any elements.

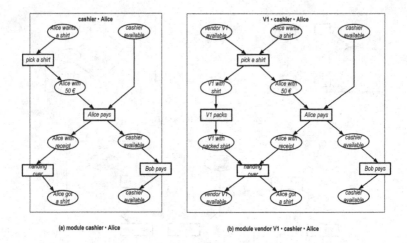

Fig. 9. Two further module compositions.

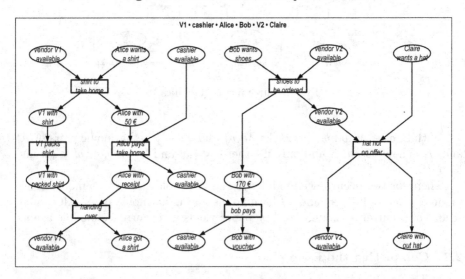

Fig. 10. Behavioral module $V := V1 \bullet cashier \bullet Alice \bullet V2 \bullet Claire \bullet Bob$.

From an abstract and more systematic point of view, the expression (3) is a bit unattractive. It would be nicer to have the module $trade := V1 \bullet cashier \bullet V2$, with all interface elements on the right, and the module $customers := Alice \bullet Claire \bullet Bob$ with all interface elements on the left. The module V in (3) is then written as $trade \bullet customers$. Indeed, this is possible without any problems, because the modules $Alice$ and $V2$ have disjoint interfaces. According to equation (2), the sequence of the two modules $V2$ and $Alice$ in (3) can be swapped.

Summing up, the module V of Fig. 10 represents a typical single *run* of a system. V provides insight into subtle details of the mutual relationship of the events of the joint behavior of the involved six agents. For example, in the presented run, the joint events of the modules $V1$, *Alice* and the *cashier* are detached from the events of the modules of the other three agents. Bob waits until the cashier is finished with Alice. But vendor $V2$ and Alice are not related at all to the cashier.

All this insight has been gained from the event log of Fig. 2, together with the intuitively obvious idea that events of the business people will never be merged, hence they come with elements in right interfaces only, and correspondingly, events of the customers will never be merged, thus all come with elements in left interfaces. The choice of left and right interface is motivated by the dichotomy between shop modules (vendors and cashiers), and client modules. Of course, right and left may be swapped here. So, the interface elements of shop modules and of client modules complement each other.

2.4 Composing an Occurrence Module from Occurrence Atoms

Here we consider an alternative way of constructing occurrence modules. In Sec. 2.3 we composed the run in Fig. 10 from the modules of the six behavioral strands of agents, given in Fig. 3. Occurrence modules are frequently, but not always, composed from modules generated by such agents. Alternatively, an occurrence module can be generated from *occurrence atoms*. An occurrence atom is a module that represents a single transition together with its surrounding arcs and places. We denote the occurrence atom of a transition t by $\underline{t}$. To correspond to the previous representation of occurrence modules, we place the left interface of an occurrence atom at the top and the right interface at the bottom of its graphical representation.

Figure 11(a), (b) and (c) show the occurrence atoms of the transitions *shirt to take home*, *V1 packs shirt*, and *Alice pays take home* of module V of Fig. 10. The composition of the three occurrence atoms in Fig. 12 is identical to the upper left part of module V. It is easy to see how the occurrence atoms of the remaining four transitions of V can be generated, and that their composition yields the entire module V. In fact, this is generally true: the occurrence atoms $\underline{t}$ of the transitions t of an occurrence net N can be arranged as a sequence $\underline{t_1}, \ldots, \underline{t_n}$ such that

$$N = \underline{t_1} \bullet \cdots \bullet \underline{t_n}. \tag{4}$$

This representation will be used in the following sections.

3 System Models

So far, we showed how to deduce a single run from a given event log. Our aim, however, is to deduce a system model from an event log. To this end, we need a manageable kind of system models. Here we derive such system models.

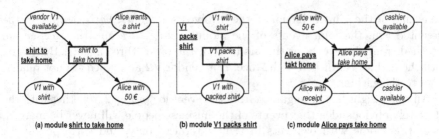

(a) module **shirt to take home**   (b) module **V1 packs shirt**   (c) module **Alice pays take home**

Fig. 11. Occurrence atoms

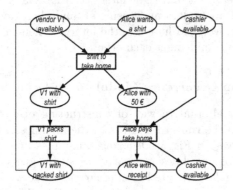

Fig. 12. Module M0: *shirt to take home* • *V1 packs shirt* • *Alice pays take home*

Many similar logs would yield many similar runs. Now we show how to extrude a system model from a set of runs. In a first step, we concentrate on the systematic management of involved data and functions.

3.1 Structures and Signatures

To cope with data and functions on data, we employ *signatures* and *signature-structures*, well-known in mathematics from general algebra and first order logic, and in informatics from algebraic specifications [12]. Figure 13 shows the signature-structure S_0 for the running example, consisting of eight sets and four functions. Each set is finite and includes real or imagined objects such as *clients*, *vendors*, *cashiers*, *products*, *vouchers*, *wrapped items*, but also more abstract items such as *money* and *descriptions of items*. In the course of systems mining, a structure like this should be provided by the provider of the logs. It may also be deducible from the logs.

A symbolic representation of a system requires abstract, symbolic representations of structures such as in Fig. 13. This is achieved by means of signatures: a *signature* Σ_0 for a structure S includes *sorted symbols*: a symbol for each set and each function of S. Figure 14 gives a signature, Σ_0, for the above structure S_0. For the sake of simple notation, for each set and each function of S_0 we write the corresponding symbol of Σ_0 in italic.

sets

M: {Alice, Bob, Clara} *set of clients*
V: {C1, C2} *set of vendors*
C: {c1} *set of cashiers*
I: {hat, shoes, shirt} *set of items*

P: {1€, 2€, ... , 1000€} *amounts of money*
V: {<hat>, <shoes>, <shirt>} *set of vouchers*
W: {[hat], [shoes], [shirt]} *set of wrapped items*
I: {hat, shoes, shirt} *set of descriptions of items*

functions

f: I ⟶ P
f(hat) = 100€   f(shoes) = 200€
                  f(shirt) = 50€

p: I ⟶ W
p(z) = [z] for z ∈ I.

v: I ⟶ V
v(z) = <z> for z ∈ I.

_: I ⟶ I
_(z) = z

set symbols	function symbols	variables
M: clients	f: I ⟶ P	x: clients
V: vendors	p: I ⟶ W	y: vendors
C: cashiers	v: I ⟶ V	z: items
I: items	_: I ⟶ I	z: description of item z
P: prices		c: cashier
V: vouchers		
W: wrapped items		
I: descriptions of items		

Fig. 13. Structure S_0, describing the system's data.

Fig. 14. Signature Σ_0 and variables for the structure S_0

Additionally, Fig. 14 shows sorted variables. Sorted symbols and variables yield *terms*, such as $f(z)$, or tuples of terms, such as $(x, f(z))$. A *valuation* β of the variables assigns to each variable v an item $\beta(v)$ of the structure S_0. For example, with $\beta(x) = Alice$, $\beta(y) = V1$ and $\beta(z) = shirt$, the tuples (y, z) and $(x, f(z))$ yield in S_0 the tuples

$$\beta(y, z) = (V1, shirt) \text{ and } \beta(x, f(z)) = (Alice, 50€) \tag{5}$$

3.2 System Atoms and Their Composition

In order to extrude a system model from a set of runs, we start from single occurrence atoms, extruding a more general model of system atoms. Figure 15 shows an example: The atom *shirt to take home* of Fig. 11(a) (repeated in Fig. 15(a)) is re-written in Fig. 15(b): information about the vendor V1, the client Alice, the item shirt, and the price 50 € moves from the module's places to its arcs. This representation is now conceived as an instantiation of the *item to take home* module in Fig. 15(c). In this module, the constant arc inscriptions of Fig. 15(b) are replaced by the variables x, y, and z, and terms $f(z)$ and z. In Figs. 15(b) and (c), the place inscriptions of the left (upper) interface places are conceived as tokens of the Petri net. Then, the firing rule of Petri nets defines the tokens for the right (lower) interface places. Figure 15(b) is now gained as the instantiation of Fig. 15(c) by means of the above valuation β as in (5). Of course, different valuations yield different instantiations of the *item to take home* system atom. This way, Fig. 15(c) is a system atom, representing many occurrence atoms.

In analogy to Fig. 15(a), (b), and (c), Fig. 15(d), (e), and (f), generalizes the *V1 packs shirt* occurrence atom as in Fig. 11(b). It is obvious how from the five remaining occurrence atoms, the corresponding system atoms can be deduced.

Figure 16 composes the seven system atoms. This is a *symbolic occurrence module*. Content wise, with the valuation β for all arc inscriptions, it is just a different representation of the occurrence module V in Fig. 10. Denotations of places and transitions have slightly been adjusted to better support intuition. The place inscriptions of V are gained in Fig. 16 by the Petri net firing rule. Further, Fig. 16 has a non-empty left and right interface, in contrast to Fig. 10.

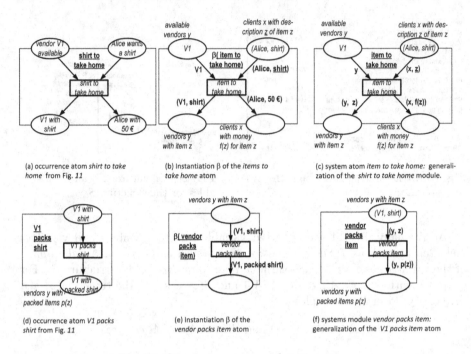

Fig. 15. Occurrence atoms and system atoms

3.3 Constructing a System Net from Symbolic Occurrence Modules

It is now possible to deduce a full-fledged Petri net model from the symbolic occurrence module of Fig. 16: Just identify equally labeled places. The resulting system net is shown in Fig. 17. The tokens of Fig. 16 are collected on the corresponding place of Fig. 17.

Figure 17 shows a high level Petri net. It specifies a lot of runs, depending on the choice of the valuation β of the variables. Furthermore, now, even when fixing the valuation β as above, each client with his description of an item may now execute any of the three events *item not on offer*, *item to be ordered* and *item to take home*, with any of the two vendors $V1$ or $V2$. This is a generalization that suggests itself from the assumptions of the system.

3.4 Deriving a Net Schema

The system model in Fig. 17 fixes the sets of vendors, clients, and items. One would prefer a specification that leaves these sets open, to be fixed as an *interpretation* of those symbols by the user of the model. For this purpose, it suggests itself to use fresh symbols, e.g. *VE*, *CL*, and *CA*, to be interpreted as sets of vendors, clients, and cashiers, as initial tokens on the places *available vendors*, *clients with descriptions of items*, and *available cashiers*, resp. However, this is not exactly what we want: An interpretation of *VE* would, for example, interpret

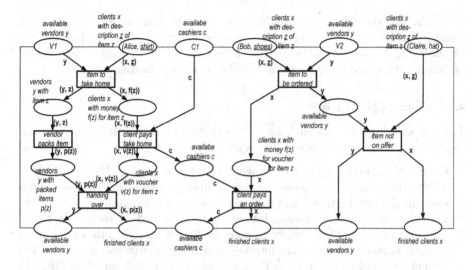

Fig. 16. Composed atoms: symbolic representation of the behavioral module V.

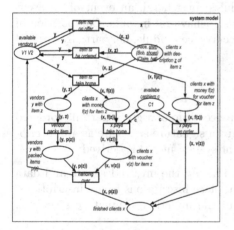

Fig. 17. System model M.

Fig. 18. Schema: symbolic initial marking.

the symbol VE by the set $\{V1, V2\}$ as one token on the place *available vendors*. Instead, we want two tokens, $V1$ and $V2$. This is represented by means of the "elm"-notation, as in the net of Fig. 18 (more details in [5]).

4 How to Mine a System Model

The notions, concepts, and constructs described in the above sections suggest to mine a system model starting from information on static system aspects such as the architecture of the system, the data structures, and the involved agents.

The data and the operations on the data are systematically represented in a signature-structure. The architecture and the agents provide the background for modeling dynamic aspects, i.e. for the derivation of occurrence modules as models for runs, and finally system modules as models for the entire system.

4.1   From Logs to Runs

The first step identifies for each agent its sequential behavior from the log, and constructs a distributed run from the agents' behaviors:

1. From a given event log, for each agent identify in the log the events which involve the agent. The sequence of those events constitute the *behavior* of the agent in the log. Figures 2 and 3 show corresponding examples.
2. Turn the behavior of each agent into an occurrence module: each event either belongs to the module's interior part, or its left or its right interface. For an element, adequate choice of the interface depends on the intended composition with elements of other modules. Figures 4(a), and (b), 5(a), and 8(a), (b), and (c) show examples.
3. Compose the agents' occurrence modules: In general, an event of an event log is involved in more than one agents' behavioral module. Composition of the modules yields a comprehensive occurrence module, i.e. partially ordered run, as in Fig. 10.

4.2   From Runs to Systems

The second step identifies for each occurrence atom of a given partially ordered run a system atom with terms over the given signature-structure as arc inscriptions. From this representation, the sought system model is derived:

4. For each occurrence atom of the run, identify the involved agents and data structures. Move this information from place inscriptions to arc inscriptions. The arc inscriptions then are terms of the underlying signature structure. Figure 15 shows examples.
5. In this representation of each occurrence atom, replace each constant symbol by a variable. This yields a system atom.
6. Compose those system atoms, as in Fig. 16.
7. In this representation, merge equally denoted places. This yields the sought system model, as in Fig. 17.
8. To achieve a purely schematic representation, replace the initial marking by a symbolic marking, as in Fig. 18.

5   Related Work

The main concepts for the theoretical foundations of process mining are based on the idea of grammar inference, grammar induction, or language identification [9], which was originally proposed by [8]. Since these theoretical models do not adequately represent all interesting aspects of business processes, a plethora of enhanced formal frameworks are developed [1]. However, none of these approaches are completely satisfactory because the role of causality, subsystems, and data are not integrated and adequately covered. Besides the theoretical work, many practical approaches originate from engineering process modeling and mining systems [7]. However, these approaches lack a theoretical foundation.

Recent work in the area of artifact-centric [4], object-centric process mining [2], and agent system mining [13], addresses these lacks already. Although these ideas clearly show improvements compared to the classical understanding of systems, models, and logs as formal languages, they still do not provide a satisfactory understanding of system architecture and the difference of abstract and concrete data structures which are strongly needed for an integrated understanding of business systems. Additionally, our understanding of an agent is rather general compared to the technical notion used by [13].

Although recent work acknowledges the need for representing causal structures, the choice is often not satisfactory. C.A. Petri formulated the concept of distributed runs as early as the late 1970s [10]. It has been taken up again and again, also under the names "true concurrency", or "partial order semantics", but initially did not prevail over sequential processes. One of the reasons for this was the comparatively complex technical apparatus for dealing with distributed processes, combined with comparatively little benefit. Meanwhile, the basic ideas of distributed runs are used in many contexts, e.g. partial order process mining [14]. Furthermore, the composition calculus, as used in this contribution, provides adequate and simple technical tools.

To cope properly with data aspects, and in particular to properly integrate behavioral and data aspects in one formal framework, we resort to signature-structures, the established formal basis of first order logic and algebraic specifications [12]. Models of really big systems are gained by composing models of subsystems. The composition calculus covers also this aspect, as developed in [5].

6   Conclusion

Classical process mining assumes a run as a sequence of events and then tries to solicit information about concurrent, independent event occurrences a and b form the observation, that in many similar logs, a and b occur in either order. We suggest to start considering a run as an unordered set of events, and then to order them, as much as reasonable, by considering agents and the composition of agents' behavior. For example, the module V of Fig. 10 provides insight into subtle details of the mutual relationship of the events of the joint behavior of the involved six agents. In the presented run, the joint events of the modules *vendor V1*, *Alice* and *cashier* are detached from the events of the modules of the other three agents. Bob waits until the cashier is finished with Alice. But vendor $V2$ and Alice are not related at all to the cashier. All this has been gained from the event log of Fig. 2, together with the intuitively obvious idea that the events of the trade components never should be merged, hence all go to the right interfaces, and correspondingly the events of the customers should never be merged, thus all go to the left interfaces. Of course, right and left may be swapped here.

In this paper, we argue that *causality, composition*, and *objects matter* while mining a system. We introduce the foundational concepts for conducting system mining. In the future, more case studies need to be done and new tools for supporting the main ideas of HERAKLIT have to be developed. So, in the future, we speculate that the two academic worlds of data and process mining will be complemented with and enhanced by systems mining allowing a deeply integrated understanding of business processes.

References

1. van der Aalst, W.: Process Mining: Data Science in Action, 2nd edn. Springer, Heidelberg (2016). https://doi.org/10.1007/978-3-662-49851-4
2. van der Aalst, W., Berti, A.: Discovering object-centric petri nets. Fund. Inform. **175**(1–4), 1–40 (2020)
3. Buijs, J.C.A.M., van Dongen, B.F., van der Aalst, W.: Quality dimensions in process discovery: the importance of fitness, precision, generalization and simplicity. Int. J. Cooper. Inf. Syst. **23**(1), 1440001/1-39 (2014)
4. Fahland, D.: Artifact-centric process mining. In: Sakr, S., Zomaya, A.Y. (eds.) Encyclopedia of Big Data Technologies, pp. 108–117. Springer, Cham (2019). https://doi.org/10.1007/978-3-319-77525-8_93
5. Fettke, P., Reisig, W.: Handbook of HERAKLIT (2021). HERAKLIT working paper, v1.1, 20 September 2021. http://www.heraklit.org
6. Fettke, P., Reisig, W.: Modelling service-oriented systems and cloud services with HERAKLIT. In: Zirpins, C., et al. (eds.) ESOCC 2020. CCIS, vol. 1360, pp. 77–89. Springer, Cham (2021). https://doi.org/10.1007/978-3-030-71906-7_7
7. Gartner: Market guide for process mining. Technical report ID G00387812 (2019)
8. Gold, E.M.: Language identification in the limit. Inf. Control **10**(5), 447–474 (1967)
9. de la Higuera, C.: Grammatical Inference: Learning Automata and Grammars. Cambridge University Press, Cambridge (2010)
10. Petri, C.A.: Non-sequential processes. Technical report ISF-77-5, Gesellschaft für Mathematik und Datenverarbeitung, St. Augustin, Federal Republic of Germany (1977)
11. Reisig, W.: Associative composition of components with double-sided interfaces. Acta Informatica **56**(3), 229–253 (2018). https://doi.org/10.1007/s00236-018-0328-7
12. Sanella, D., Tarlecki, A.: Foundations of Algebraic Specification and Formal Software Development. Springer, Heidelberg (2012). https://doi.org/10.1007/978-3-642-17336-3
13. Tour, A., Polyvyanyy, A., Kalenkova, A.: Agent system mining: vision, benefits, and challenges. IEEE Access **9**, 99480–99494 (2021)
14. van der Aa, H., Leopold, H., Weidlich, M.: Partial order resolution of event logs for process conformance checking. Decis. Support Syst. **136**, 113347 (2020)

Conformance Checking
over Stochastically Known Logs

Eli Bogdanov[1], Izack Cohen[2(✉)], and Avigdor Gal[1]

[1] Faculty of Industrial Engineering and Management,
Technion – Israel Institute of Technology, 3200003 Haifa, Israel
avigal@technion.ac.il
[2] Faculty of Engineering, Bar-Ilan University, 5290002 Ramat Gan, Israel
izack.cohen@biu.ac.il
https://izackcohen.com, https://agp.iem.technion.ac.il/avigal/

Abstract. With the growing number of devices, sensors and digital systems, data logs may become uncertain due to, *e.g.*, sensor reading inaccuracies or incorrect interpretation of readings by processing programs. At times, such uncertainties can be captured stochastically, especially when using probabilistic data classification models. In this work we focus on conformance checking, which compares a process model with an event log, when event logs are stochastically known. Building on existing alignment-based conformance checking fundamentals, we mathematically define a stochastic trace model, a stochastic synchronous product, and a cost function that reflects the uncertainty of events in a log. Then, we search for an optimal alignment over the reachability graph of the stochastic synchronous product for finding an optimal alignment between a model and a stochastic process observation. Via structured experiments with two well-known process mining benchmarks, we explore the behavior of the suggested stochastic conformance checking approach and compare it to a standard alignment-based approach as well as to an approach that creates a lower bound on performance. We envision the proposed stochastic conformance checking approach as a viable process mining component for future analysis of stochastic event logs.

1 Introduction

Process mining relies on data that are typically stored in the form of event logs and collections of traces where each trace is a sequence of events and activities that were created following a process realization. Process mining tasks, such as conformance checking, use event logs to achieve their goal (*e.g.*, assessing to what degree a process model and an event log conform) of improving the process model that generates these logs.

The fourth industrial revolution [1], which is bridging our digital and physical worlds, is producing an abundance of event data from multiple sources such as social media networks [2], sensors located within smart cities (*e.g.*, the 'Green Wall' project in Tel Aviv and Nanjing), medical devices and much more.

© Springer Nature Switzerland AG 2022
C. Di Ciccio et al. (Eds.): BPM 2022, LNBIP 458, pp. 105–119, 2022.
https://doi.org/10.1007/978-3-031-16171-1_7

Differently from data within traditional information systems, these data may involve uncertainty due to technical reasons such as sensor inaccuracy, the use of probabilistic data classification models, data quality reduction during processing, and low quality of data capturing devices. Human generated data may be uncertain as well, due to fake news and mediator interventions.

In this work, we focus on process mining with Stochastically known (SK) event data [3] where the probability distribution functions of the event data are known.[1] By way of motivation, consider a use-case of food preparation processes, captured in video clips that are analyzed by a pre-trained Convolutional Neural Network (CNN) to predict activity classes and their sequence within an observed video. To extract the trace of the realized process, one can use the softmax layer of the CNN to yield a discrete probability distribution of the predicted activity classes in the observed video. This probabilistic knowledge, in turn, can serve as a basis for an SK log. Specifically, we develop a conformance checking algorithm over SK data. Building on existing alignment-based conformance checking fundamentals, we mathematically define a stochastic trace model, a stochastic synchronous product, and a cost function that reflects the uncertainty of events in a log. Then, we search for an optimal alignment over the reachability graph of the stochastic synchronous product to find an optimal alignment. The main contributions of this work are:

1. We characterize and mathematically define the building blocks for stochastic conformance checking, including a stochastic trace model and a stochastic synchronous product.
2. We develop a novel conformance checking algorithm between a model and an SK trace.
3. Using publicly available data sets, we evaluate the performance of stochastic conformance checking and highlight unique features of our proposed algorithm.

The rest of the paper is organized as follows. In Sect. 2, we develop the model followed by presentation of our stochastic alignment algorithm (Sect. 3). Empirical evaluation of the two is given in detail in Sect. 4. The related literature is presented in Sect. 5 and the final section (Sect. 6) concludes the paper and offers directions for future research.

2   Stochastic Trace Model

Uncertain data have recently become a subject of interest among the process mining community [4–6]. Table 1 [3] presents a model/observation classification scheme that is based on the number of models present in a log and whether the log is deterministically or stochastically known. In this work we focus on Case 5, handling a Deterministically known (DK) process model and an SK trace,

---

[1] It is also denoted as 'weakly uncertain' event data in the process mining literature; see [4].

where the decision-maker wishes to identify a conformance measure between the process and the SK trace. While the suggested approach can be extended to solve Case 7, we leave this extension as well as other cases for future work.

Table 1. Eight cases according to the characteristics of the process and observed log, from Cohen and Gal [3]. The present paper focuses on Case 5 (highlighted).

Model (Data set) → ↓ Observation (Log)	Single process		Multiple processes	
	DK	SK	DK	SK
Deterministically Known (DK)	1	2	3	4
Stochastically Known (SK)	5	6	7	8

Following Cohen and Gal [3], we use DK to describe a given and known process or event log, which is the common setting in the process mining literature. An SK event log has at least one event attribute that can be characterized via a probability distribution. Table 2 illustrates an SK trace, which we use as the running example throughout the paper.

Table 2. SK data, which is aligned with Case 5 in Table 1 in [3].

Case ID	Event ID	Activity	Timestamp
1	e_1	$\{A : 1.0\}$	13-08-2020T12:00
1	e_2	$\{B : 0.2, C : 0.8\}$	13-08-2020T14:55
1	e_3	$\{D : 0.6, E : 0.2, F : 0.1, G : 0.1\}$	15-08-2020T17:39
1	e_4	$\{F : 1.0\}$	15-08-2020T19:47

We now introduce our primary notation and related definitions. We consider a finite set of activities $\mathcal{A}$ and a Petri net N with initial and final markings m_i and m_f, respectively. The Petri net is composed of finite sets of places P, transitions T and flow relations F, which are directed edges among places and transitions. Each transition is associated with an activity $a \in \mathcal{A} \cup \tau$ by the labeling function $\lambda : T \rightarrow A^\tau$ ($A^\tau \equiv \mathcal{A} \cup \tau$). τ is a silent activity separate from the other activities in $\mathcal{A}$. Differently from a DK trace that includes a sequence of activities with probability 1, the activities in an SK trace are associated with a probability function (*e.g.*, the next transition may be 'act1' with probability p or 'act2' with probability $1 - p$). We reflect the stochastic nature of the traces using a weight function $W : T \rightarrow (0, 1)$ that assigns a firing probability to each transition.

Our modeling approach is inspired by a conformance checking algorithm [7] (pp. 125–158) to align a DK trace and a model's execution sequence such that the cost of dissimilarities is minimized. The algorithm by [7] cannot be used directly with SK traces. Our proposed model, however, aims to provide this ability. In what follows, we assume prior knowledge about alignment-based conformance checking and related definitions (*e.g.*, system net, process and trace models, and synchronous product). We refer interested readers to [7] for a thorough description of relevant definitions and methods. We start by defining a stochastic trace model.

Definition 1 (Stochastic Trace Model). *Let* $A \subseteq \mathcal{A}$ *be a set of activities, and* $\sigma \in A^*$ *a sequence over these activities. A stochastic trace model,* $STN = ((P, T, F, \lambda, W), m_i, m_f)$ *is a system net such that* $P = \{p_0, ..., p_{|\sigma|}\}$, $T \in \{t_{11}, ..., t_{|\sigma|n_\sigma}\}$, $F \subseteq (P \times T) \cup (T \times P)$ *and* $W : T \to (0, 1) \mid \sum_{j=1}^{n_i} W(t_{ij}) = 1$, $\forall 1 \leq i \leq |\sigma|$ *where* n_i *is the number of alternative transitions between place* p_{i-1} *and* p_i. $W(t_{i\cdot})$ *is a probability function assigning to each alternative transition* j *a firing probability. Additionally, let* $m_i = [p_0]$ *and* $m_f = [p_{|\sigma|}]$.

Figure 1 offers a visual illustration of a stochastic trace model for our running example from Table 2, where transition t_{11} is activity A, t_{21} and t_{22} are activities B and C, respectively, and so on. The stochastic trace model generalizes a trace model by allowing a place i to have multiple incoming and outgoing edges denoted by j, which lead to and from alternative transitions. Each transition has a single outgoing edge from a place and a single incoming edge to a place. Additionally, each transition is associated with a firing probability. For each two places in the Petri net, the sum of firing probabilities of their alternative transitions is 1.

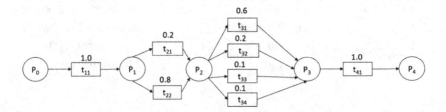

Fig. 1. Stochastic trace model illustration

3   Stochastic Alignment Algorithm

A synchronous product combines process and trace models such that each pair of transitions that are labeled with the same activity are denoted a synchronous transition. Nonsynchronous transitions are represented by pairing an activity with >> and are associated with a cost of 1. An optimal alignment between a trace and a model is the execution sequence of the model for which the alignment between the trace and the sequence has the lowest possible cost. De facto, this is an execution sequence of the synchronous product model that produces the lowest cost.

While deterministic traces have a single execution sequence, for SK traces a synchronous product procedure should align multiple model execution sequences with multiple trace execution sequences. We search for the optimal alignment using the reachability graph of the synchronous product. Towards this end, we need to extend the standard version of a synchronous product by including probability functions that capture the SK nature of the trace. The probability functions assign a firing probability to each synchronous move of the trace and the model. The probability of the synchronous move is equal to the probability of the same transition in the stochastic trace model as defined next.

Definition 2 (Stochastic Synchronous Product).
Let

$$SN = ((P^{SN}, T^{SN}, F^{SN}, \lambda^{SN}), m_i^{SN}, m_f^{SN})$$

be a process model and

$$STN = ((P^{STN}, T^{STN}, F^{STN}, \lambda^{STN}, W^{STN}), m_i^{STN}, m_f^{STN})$$

a stochastic trace model. The stochastic synchronous product $SSN = ((P, T, F, \lambda, W), m_i, m_f)$ *is a system net such that:*

- $P = P^{SN} \cup P^{STN}$ *is the set of places,*
- $T = T^{MM} \cup T^{LM} \cup T^{SM} \subseteq (T^{SN} \cup \{>>\}) \times (T^{STN} \cup \{>>\})$ *is the set of transitions where* $>>$ *denotes an SSN transition in which either the model or the trace executes an activity and its counterpart does not, i.e.,* $>> \notin T^{SN} \cup T^{STN}$, *with*
  $T^{MM} = T^{SN} \times \{>>\}$ *(model moves),*
  $T^{LM} = \{>>\} \times T^{STN}$ *(log moves), and*
  $T^{SM} = \{(t_i, t_j) \in T^{SN} \times T^{STN} \mid \lambda^{SN}(t_i) = \lambda^{STN}(t_j)\}$ *(synchronous moves).*
- $F = \{(p, (t_i, t_j)) \in P \times T \mid (p, t_i) \in F^{SN} \vee (p, t_j) \in F^{STN}\} \cup \{((t_i, t_j), p) \in T \times P \mid (t_i, p) \in F^{SN} \vee (t_j, p) \in F^{STN}\}$,
- $m_i = m_i^{SN} + m_i^{STN}$,
- $m_f = m_f^{SN} + m_f^{STN}$ *and,*
- $\forall (t_i, t_j) \in T$ *it holds that* $\lambda((t_i, t_j)) = (l_i, l_j)$, *where* $l_i = \lambda^{SN}(t_i)$ *if* $t_i \in T^{SN}$, *and* $l_i => >$ *otherwise; and* $l_j = \lambda^{STN}(t_j)$, *if* $t_j \in T^{STN}$, *and* $l_j => >$ *otherwise. Finally,*
- *the probability function* $W : T \to (0, 1) \mid W^{SSN}(t_i, t_j) = W^{STN}(t_j), \forall (t_i, t_j) \in T^{SSN} : \lambda^{SN}(t_i) = \lambda^{STN}(t_j)$ *assigns firing probabilities to transitions of synchronous moves.*

The stochastic synchronous product is a combination of a process model that may yield multiple execution sequences (traces) and a stochastically known trace model that is noisy. Thus, the 'real' deterministic trace can be only deduced with probability. The transitions of the stochastic synchronous product are a union of synchronous and nonsynchronous transitions. To combine a process model and a trace in a system net that represents the synchronous product, each pair of transitions that are labeled with the same activity is added as a synchronous transition. Nonsynchronous transitions, which include a process (trace) activity that cannot be matched with the same activity on the trace (model), are paired with >>. Figure 2 illustrates the stochastic synchronous product of a model (its starting place is P_{01}) and the stochastic trace of our running example (its starting place is P_{02}). The first transition in both the model and the trace is given the label "activity A" and thus, a new synchronous transition is created—namely, transition (A, A). The original transitions both in the model and the trace are paired with the symbol >> and are added to the new net as well.

We are now ready to introduce our algorithm, $S\text{-}ABCC$ (Stochastic Alignment-Based Conformance Checking), as a solution to the problem of finding the lowest-cost execution sequence of the synchronous product. We observe that this is equivalent to finding the shortest path over the synchronous product's reachability graph, where the sum of costs across path edges is the total path length.

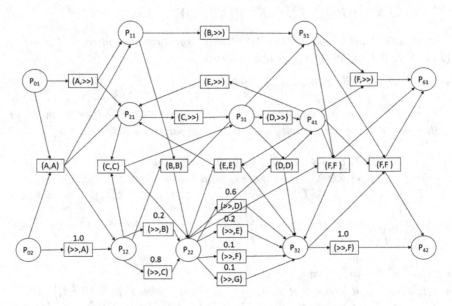

Fig. 2. Stochastic synchronous product illustration

Given an initial marking m_i of a stochastic synchronous process model SSN, we denote the corresponding system net as $N = (P, T, F, \lambda, W)$ and its set of reachable markings as $RS(N)$. The reachability graph of N, denoted by $RG(N)$, is a graph in which the set of nodes is the set of markings $RS(N)$ and the edges correspond to firing transitions, where each edge in $RG(N)$ corresponds to a transition of the stochastic synchronous process SSN. Formally, an edge $(m_1, t, m_2) \in RS(N) \times T \times RS(N)$ exists, if and only if $m_1[t\rangle m_2$. The shortest path from the initial to the final marking in $RG(N)$ corresponds to the lowest-cost execution sequence of SSN. We model the transition probabilities of the

SK trace in the reachability graph by assigning weights (costs) to the edges as discussed next.

Recall that SSN is the stochastic synchronous product of $SN = ((P^{SN}, T^{SN}, F^{SN}, \lambda^{SN}), m_i^{SN}, m_f^{SN})$ and a stochastic trace $STN = ((P^{TN}, T^{TN}, F^{TN}, \lambda^{TN}, W^{TN}), m_i^{TN}, m_f^{TN})$. For every synchronous move, transition $t' = (t_i, t_j)$ in SSN and its corresponding edge e' in $RG(N)$, the cost of e' is calculated by

$$Weight(e') = 1 - e^{1 - \frac{1}{W(t')}}, \quad \forall t' = (t_i, t_j) \in T^{SSN} \mid \lambda^{SN}(t_i) = \lambda^{STN}(t_j) \quad (1)$$

where $W(t')$ is the firing probability of transition t', and 1 otherwise ($W(e') = 1$, $\forall t' = (t_1, t_2) \in T^{SSN} \mid t' \in T^{SN} \times \{>>\} \vee \{>>\} \times T^{STN}$ (model moves or log moves, respectively)).

The cost function (Eq. 1) transforms firing probabilities into costs. We use a monotonic, non-linear cost function such that each edge e' in the reachability graph $RG(N)$ satisfies the following: $0 \le Weight(e') \le 1$. The monotonicity property assures that if the probability assigned to a transition t is higher than the probability of its alternative transition t', $W(t) > W(t')$, it follows that for the corresponding edges in the reachability graph (for synchronous moves) e and e' satisfy the following: $Weight(e) < Weight(e')$. In other words, the penalty on transitions participating in synchronous moves is higher as they are more uncertain. Before choosing the cost function, we experimented with other functions that satisfy the monotonicity property such as $Weight(e') = -ln(W(t'))$ and $Weight(e') = 1 - W(t')$. The following property (which proof is omitted) bounds the costs of synchronous moves.

Property 1. The cost function (Eq. 1), $f(x) = 1 - e^{1 - \frac{1}{x}}$, satisfies the following properties for synchronous moves:

1. The cost of an edge in $RG(N)$ approaches 0^+ as the firing probability of its transition approaches 1,
2. it approaches 1 as the firing probability of the transition approaches 0, and
3. $1 \le f(x) \le 0, \ \forall x \in (0, 1]$.

For the deterministic setting, the cost of each edge in $RG(N)$ is either 0 or 1 and thus, the deterministic setting can be seen as a special case of our setting with the firing probability of each transition set to 1. Given a stochastic synchronous product SSN (Definition 2) and the cost function (Eq. 1), any shortest path algorithm (*e.g.*, Dijkstra [7]) can be applied to find the shortest (cheapest) path from the initial to the final markings – this path corresponds to an optimal alignment between the stochastic trace and the model. To illustrate, Fig. 3 presents the reachability graph of the stochastic synchronous product in Fig. 2 and the shortest path.

4   Empirical Evaluation

We evaluate *S-ABCC* against a standard alignment-based conformance and a lower bound on the conformance cost [4]. We start with a description of the benchmark data sets (Sect. 4.1), followed by an explanation of the experiment design (Sect. 4.2). We report on the outcome of the empirical evaluation in Sect. 4.3.

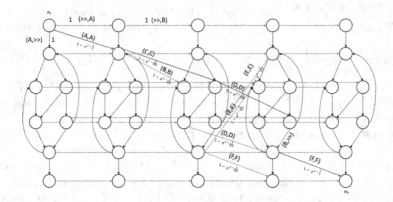

Fig. 3. The reachability graph of the stochastic synchronous product in Fig. 2. The red edges mark the optimal path after applying the Dijkstra algorithm. (Color figure online)

4.1   The Datasets

We used two publicly available real-world datasets as a baseline for our experiments: BPI 2019 and BPI 2012. The BPI 2019 data set contains over 1.5 million events for purchase orders that were collected from a large international coatings and paints company in the Netherlands. The dataset consists of over 250,000 traces relating to 42 activities performed by 627 users. The BPI 2012 dataset consists of about 262,000 events and 13,000 applications for personal loans or overdraft approvals held by a Dutch financial institute.

4.2   Data Preparation and Experiment Design

For each of the data sets, we discovered a baseline model using 15 randomly chosen traces via the Inductive Miner (IM) algorithm and the PM4PY package.

Stochastic traces were generated from traces that were not utilized for model discovery. We used 100 traces—15 for the model discovery while the remaining 85 were transformed into stochastic traces. The transformation procedure iterates over each trace, adding alternative transitions with random activities. Both original and added transitions are assigned a firing probability. For example, if the original log contained the following record: $\{CaseID : 1, EventId : e1, Activity : A\}$, a possible corresponding stochastic record after adding transitions with random activities and firing probabilities is $\{CaseID : 1, EventId : e1, Activity : [A : 0.4, B : 0.4, C : 0.2]\}$.

We control the following parameters when preparing the stochastic traces.

- Number of alternative transitions, N_t, varied between 2 and 4. Consider, for example, $N_t = 2$, which is two alternative transitions for trace $<A, B, C>$. Then for each of the three events, a second alternative transition is added with an activity that is randomly chosen from the set of activities.
- Value of the firing probability assigned to the original transition in each set of alternative transitions, P_f. This parameter is set to one of three values, $P_f \in (0.55, 0.75, 0.95)$. Since the sum of firing probabilities across each set of alternative transitions equals 1, the leftover probability, $1 - P_f$, is randomly split between the other alternative transitions.
- Portion of the uncertain traces, T_p. When $T_p = 0$, the considered trace is deterministic. We increased the parameter's value in steps of 0.05. For each iteration in which we increased T_p, we selected all the traces from the previous iteration and randomly selected 5% of each trace transitions to be transformed into alternative transitions. The selected 5% only included events without alternative transitions to ensure that when $T_p = 1$, 100% of the trace events would have alternative transitions.

We note, in passing, that the stochastic traces that we generated resemble the stochastic output of neural networks for classifying activities in video clips or of sensors for identifying observed signals (for more information, refer to [3]).

4.3 Results

Figure 4 demonstrates the sensitivity of the suggested approach to the distribution of the firing probabilities in the sense that changing the firing probability affects the average conformance cost. Specifically, conformance cost decreases with P_f as we get closer to the deterministic setting until it hits the red '+' marker in Fig. 4 in which $P_f = 1$. In fact, the suggested model accommodates the deterministic setting in the sense that when assigning $P_f = 1$, the suggested model generates the same conformance cost as does conventional alignment-based conformance checking. Under the suggested model, the optimal alignment carries additional conformance costs compared to its deterministic counterpart due to uncertainty. In a deterministic setting, synchronous moves do not induce a cost, which makes sense since there is only a single trace path. Under an SK setting, synchronous moves are associated with a non-negative cost due to uncertainty on the trace path. The extra cost embodies the level of uncertainty for each possible trace realization. We can argue that not accounting for uncertainty costs would lead to a situation in which as the level of uncertainty increases (*e.g.*, by having more transitions in parallel), the number of possible trace realizations grows and thus we have a greater chance of finding a better conforming trace that is associated with lower conformance costs. This situation is undesirable unless we are seeking a lower bound on the conformance cost (see [4]).

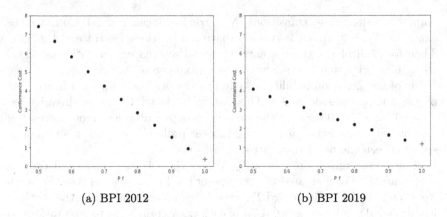

(a) BPI 2012   (b) BPI 2019

Fig. 4. Average conformance cost as as a function of the firing probability, P_f of the original trace transition. We set $T_p = 1$, where each event in the original trace included 2–4 alternative transitions – $N_t \in (2, 3, 4)$. The '+' marker corresponds to a deterministic setting. (Color figure online)

Figure 5 presents the conformance cost as a function of the stochastic trace portion size for the BPI 2012 data set (results for BPI 2019 showed similar tendencies and are not included due to space considerations). Inspired by Pegoraro, Uysal, and Van Der Aalst [4], the original traces were modified prior to adding alternative transitions in one of four ways by: 1) randomly altering the activity label for 30% of the events; 2) randomly swapping 30% of the events with either their successor or predecessor where first and last events in a trace were only swapped with their successor and predecessor, respectively; 3) randomly duplicating 30% of the trace events; and 4) all of the above modifications. After applying a modification, we turn back to the general preprocessing procedure of iteratively adding alternative transitions as detailed in Sect. 4.2. It can be seen in Fig. 5 that the conformance cost of the SK traces increases with T_p. On the other hand, the conformance cost of the lower bound, which does not account for probabilities, decreases with T_p. This occurs because higher T_p values imply more possible traces and thus additional alignment opportunities while the lower bound does not consider the realization probability of these traces. The result is that the gap, in conformance costs, between the lower bound and the suggested approach that acknowledge uncertainty increases with T_p.

Next, we evaluated the conformance cost of traces with different lengths. For this, the traces were sorted into groups according to their length, so that group 1 contains traces with a length of 0–9, group 2 contains traces with a length of 10–29 and so on. Following this, we randomly chose three traces from each group (a total of 15 traces) and discovered a model from these traces. Each data point in Fig. 6 represents the average conformance cost of all the traces that were used

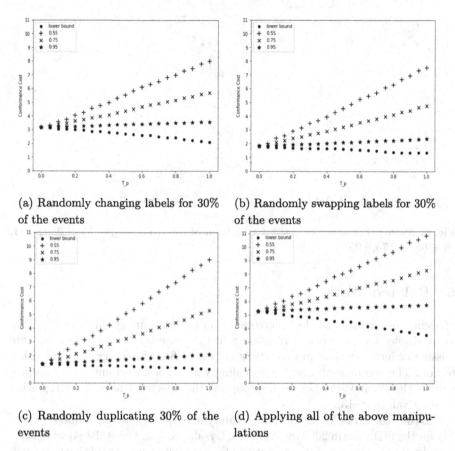

(a) Randomly changing labels for 30% of the events

(b) Randomly swapping labels for 30% of the events

(c) Randomly duplicating 30% of the events

(d) Applying all of the above manipulations

Fig. 5. Average conformance cost as a function of T_p, the trace portion with alternative transitions for the four preprocessing modifications as evaluated for the BPI 2012 data set. Different types of markers denote different P_f values and the lower bound; $N_t = 2$

for the evaluation, i.e., all the traces within a group excluding the traces that were used for the model discovery. Figure 6 demonstrates that the conformance cost is increasing with the trace length (apart from the lower bound, for the same reasons explained earlier). The observed behavior follows from the fact that longer stochastic traces have a higher number of possible realizations, which may possibly lead to a better alignment, compared to shorter ones since the number of realizations of a stochastic trace with $T_p = 1$ and $N_t = 2$ is 2^n where n is the length of the trace. We note that the additional cost from synchronous moves outweighs, on average, the reduced cost that may result from a better alignment.

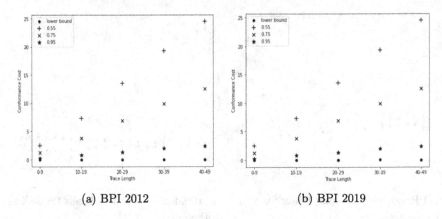

(a) BPI 2012   (b) BPI 2019

Fig. 6. Average conformance cost as a function of the trace length; $N_t = 2$, $T_p = 1$, $P_f \in (0.55, 0.75, 0.95)$

5   Related Work

Modeling uncertainty has been introduced in process mining only recently. Previous studies focused on uncertain data in the sense that some of the data are missing or incorrect and uncertainty is not quantified via any probability distribution. The common approach for dealing with such uncertainty is by preprocessing the event log either by filtering out the affected traces or by repairing existing values [8–13].

To the best of our knowledge, uncertainty in event logs was introduced explicitly for the first time in 2020 by Pegoraro, Uysal, and Van Der Aalst [4] who introduced a new taxonomy of uncertainty on the attribute level. At this level, the values of the event attributes are not missing or incorrect but rather appear as a set of possible values and in some cases, the likelihood of each possible value is known or could be estimated. The authors defined two types of uncertainty—namely *strong uncertainty* and *weak uncertainty*. The former relates to unknown probabilities between the possible values for the attribute while the latter assumes complete probabilistic knowledge in the form of a probability distribution. The strong uncertainty setting has been addressed in multiple works. A conformance checking technique was proposed by [5] to compute a lower bound on the conformance cost. Pegoraro, Uysal, and Van Der Aalst [6] described a discovery technique based on uncertain logs that represent an underlying process. In [14] and [15], the authors proposed an efficient way to construct behavior graphs, which are a graphical representation of precedence relationships among events, for logs with strong uncertain data. By using these graphs, one can discover models from logs through methods based on directly-follows relationships such as the inductive miner [6]. Van der Aa, Leopold, and Reijers [16] suggested a way to calculate the probability of a trace with an uncertain event-to-activity mapping. In a recent work, Bergami et al. [17], the authors suggested a technique to compute conformance cost for a probabilistic discovered model and deterministic traces. This work is the first to tackle the problem of conformance checking with SK logs.

6   Conclusion and Future Work

We developed a conformance checking model for an SK trace in which probability distribution functions are given. Such a setting may characterize situations in which data logs originate from sensors or probabilistic models. Differently from other conformance checking models, ours explicitly considers the probability values and at the same time accommodates standard (deterministic) alignment-based conformance checking.

When constructing the $S\text{-}ABCC$, in favor of model development, we defined a stochastic trace model and a stochastic synchronous product. Using the stochastic synchronous product and its set of reachable markings, we constructed the corresponding reachability graph. By formulating a bounded non-linear cost function that takes the firing probability as an input, we assigned costs to the edges of the reachability graph that correspond to the stochastic synchronous product. In a final step, we searched over the graph for the shortest (cheapest) path, which represents an optimal alignment where the cost is the conformance cost. Via structured experiments with two well-known benchmarks, we analyzed the characteristics of $S\text{-}ABCC$ and compared it to the deterministic alignment-based conformance checking approach and to a lower bound on the conformance cost. On average, the conformance cost of the stochastically known traces converges to their deterministic counterparts as the firing probabilities approach 1. As expected, lower values of firing probability that imply higher uncertainty correspond to higher conformance costs for the same traces. This phenomenon is confirmed when the uncertainty increases due to larger uncertain trace portions. Finally, we observed that conformance costs tend to be higher for longer stochastic traces compared to shorter ones. This occurs because, in general, longer traces may include more synchronous moves that have non-negative costs in the stochastic settings.

This work opens up several interesting future research directions. The first is to use the suggested conformance checking approach to restore the most likely realization from SK traces. Possible applications may include improving the accuracy of machine learning classifiers and cleaning errors in datasets. Another direction is to find both upper and lower bounds on conformance cost. Finally, it is worth exploring how different cost functions and search algorithms may affect the performance of $S\text{-}ABCC$.

Acknowledgements. This research was supported by THE ISRAEL SCIENCE FOUNDATION grants No. 1825/20 (AG) and 226/21 (IC).

References

1. Schwab, K.: The fourth industrial revolution. Currency (2017)
2. Sener, F., Yao, A.: Unsupervised learning and segmentation of complex activities from video. In: Proceedings of the IEEE Conference on Computer Vision and Pattern Recognition, pp. 8368–8376 (2018)
3. Cohen, I., Gal, A.: Uncertain process data with probabilistic knowledge: problem characterization and challenges. In: Proceedings of the International Workshop Problems21, Co-located with the 19th International Conference on Business Process Management BPM 2021, Italy, Published in CEUR Workshop Proceedings, vol. 2938, pp. 51–56 (2021)
4. Pegoraro, M., Uysal, M.S., Van Der Aalst, W.: Conformance checking over uncertain event data. arXiv Preprint arXiv:2009.14452 (2020)
5. Pegoraro, M., van der Aalst, W.: Mining uncertain event data in process mining. In: 2019 International Conference on Process Mining (ICPM), pp. 89–96. IEEE (2019)
6. Pegoraro, M., Uysal, M.S., van der Aalst, W.M.P.: Discovering process models from uncertain event data. In: Di Francescomarino, C., Dijkman, R., Zdun, U. (eds.) BPM 2019. LNBIP, vol. 362, pp. 238–249. Springer, Cham (2019). https://doi.org/10.1007/978-3-030-37453-2_20
7. Carmona, J., van Dongen, B., Solti, A., Weidlich, M.: Conformance Checking: Relating Processes and Models. Springer, Cham (2018). https://doi.org/10.1007/978-3-319-99414-7
8. Suriadi, R.A., Ter Hofstede, A.H.M., Wynn, M.T.: Event log imperfection patterns for process mining: towards a systematic approach to cleaning event logs. Inf. Syst. **64**, 132–150 (2017)
9. Wang, J., Song, S., Lin, X., Zhu, X., Pei, J.: Cleaning structured event logs: a graph repair approach. In: 2015 IEEE 31st International Conference on Data Engineering, pp. 30–41. IEEE (2015)
10. Conforti, R., La Rosa, M., ter Hofstede, A.H.M.: Filtering out infrequent behavior from business process event logs. IEEE Trans. Knowl. Data Eng. **29**(2), 300–314 (2016)
11. Sani, M.F., van Zelst, S.J., van der Aalst, W.M.P.: Improving process discovery results by filtering outliers using conditional behavioural probabilities. In: Teniente, E., Weidlich, M. (eds.) BPM 2017. LNBIP, vol. 308, pp. 216–229. Springer, Cham (2018). https://doi.org/10.1007/978-3-319-74030-0_16
12. van Zelst, S.J., Fani Sani, M., Ostovar, A., Conforti, R., La Rosa, M.: Filtering spurious events from event streams of business processes. In: Krogstie, J., Reijers, H.A. (eds.) CAiSE 2018. LNCS, vol. 10816, pp. 35–52. Springer, Filtering spurious events from event streams of business processes (2018). https://doi.org/10.1007/978-3-319-91563-0_3
13. Conforti, R., La Rosa, M., ter Hofstede, A.H.M.: Timestamp repair for business process event logs. https://minerva-access.unimelb.edu.au/handle/11343/209011 (2018)
14. Pegoraro, M., Uysal, M.S., van der Aalst, W.M.P.: Efficient construction of behavior graphs for uncertain event data. In: Abramowicz, W., Klein, G. (eds.) BIS 2020. LNBIP, vol. 389, pp. 76–88. Springer, Cham (2020). https://doi.org/10.1007/978-3-030-53337-3_6
15. Pegoraro, M., Uysal, M.S., Van Der Aalst, W.: Efficient time and space representation of uncertain event data. Algorithms **13**(11), 285 (2020)

16. Van der Aa, H., Leopold, H., Reijers, H.A.: Efficient process conformance checking on the basis of uncertain event-to-activity mappings. IEEE Trans. Knowl. Data Eng. **32**(5), 927–940 (2019)

17. Bergami, G., Maggi, F.M., Montali, M., Peñaloza, R.: A tool for computing probabilistic trace alignments. In: Nurcan, S., Korthaus, A. (eds.) CAiSE 2021. LNBIP, vol. 424, pp. 118–126. Springer, Cham (2021). https://doi.org/10.1007/978-3-030-79108-7_14

Alpha Precision: Estimating the Significant System Behavior in a Model

Benoît Depaire[1] , Gert Janssenswillen[1](✉) , and Sander J.J. Leemans[2]

[1] Business Informatics, Hasselt University, Hasselt, Belgium
{benoit.depaire,gert.janssenswillen}@uhasselt.be
[2] Queensland University of Technology, Brisbane, Australia
s.leemans@qut.edu.au

Abstract. One of the goals of process discovery is to construct, from a given event log, a process model which correctly represents the underlying system. As with any abstraction, one does not necessarily want to represent all possible behavior, but only the significant behavior. While various discovery algorithms support this use case of discovering the significant process behavior, proper evaluation measures for this use case appear to be missing.

Therefore, this paper presents a new precision metric that quantifies to what extent the discovered model contains significant system behavior. Besides being a metric with a clear and intuitive interpretation, the metric distinguishes itself in two other areas. Firstly, it introduces the concept of α-significance, which only measures precision with respect to significant behavior. Secondly, it is designed as a system measure and estimates the precision with respect to the underlying system rather than the observed log. This work introduces a new precision measure and a statistical estimation method. Additionally, an empirical demonstration and evaluation of the metric are provided, which creates initial insights and knowledge about the performance and characteristics of the new measure. The results show that the α-precision measure provides a solid foundation for future work on developing quality measures for this particular use case.

Keywords: Process discovery · Precision · Stochastic process models

1 Introduction

Various information systems increasingly support current business processes, and create a digital trail of process execution information. These digital trails can be transformed into an event log, which records at a minimum the executed activities and their order for each case. Given such event logs, the goal of process discovery is then to discover a model representing the underlying process (also called system) as closely as possible from the event log.

Event logs are only a sample of the possible process or system behavior. Therefore, most process discovery algorithms try to generalize the observed

© Springer Nature Switzerland AG 2022
C. Di Ciccio et al. (Eds.): BPM 2022, LNBIP 458, pp. 120–136, 2022.
https://doi.org/10.1007/978-3-031-16171-1_8

behavior to capture the whole system behavior rather than the log behavior only. At the same time, a system can contain a large amount of infrequent behavior and trying to represent all this behavior in a single (visual) model quickly results in non-interpretable spaghetti models.

Hence, we focus on the particular use case where one wants to rediscover only the system's significant—typical—behavior. Fortunately, various process discovery algorithms exist that contain mechanisms and parameters that support this use case. That is, discovery algorithms have introduced different ways to classify and filter insignificant behavior: [18] classifies traces of the log as insignificant if they traverse little-used parts of an intermediate behavior abstraction; [7,8] classify little-used model edges as insignificant; [15,22] classify edges of a behavior abstraction as insignificant based on frequency; [2,16] search for a most likely model, thereby implicitly classifying behavior that does not fit that intermediate result as insignificant; and [7] hides insignificant details in hierarchy.

Following the model-log-system quality paradigm in process mining [3], two criteria exist to evaluate the quality of a process model against the system, model-system fitness, and model-system precision. This paper focuses on model-system precision, which quantifies to what extent the process model only contains system behavior. Unfortunately, the existing precision measures fall short of the presented use case for three reasons.

Firstly, existing precision measures do not distinguish between significant (typical) and insignificant (infrequent) process behavior. Consequently, a model that contains a lot of insignificant behavior is still considered to be very precise by these measures, as long as that insignificant behavior is part of the system or log.

Secondly, most precision measures are developed as model-log measures. Consequently, they do not measure to what extent the model only contains behavior from the system, but rather quantify to what extent the model only contains behavior observed in the log. Research has also shown that these model-log measures have limited value when used as proxies for model-system measures [9].

Thirdly, many quality measures in process mining became so advanced over time that an unambiguous interpretation of the precision value is no longer possible. For many measures, the precision value has become a number that is the result of complex computation. While it still correlates to the precision of the model, it lacks a meaningful and unambiguous interpretation.

This research aims to design and introduce a new precision measure that tackles these limitations and (indirectly) supports the use case of discovering significant system behavior. The paper makes three main contributions:

- A first-of-its-kind precision measure is introduced, quantifying the amount of significant behavior in a process model and providing measurement values that have a meaningful and unambiguous interpretation.
- A statistical method based on Bayesian Inference is provided to efficiently estimate the system precision based on a given event log.
- Initial empirical insights into the performance of this new precision measure are provided, which opens up avenues for follow-up research.

The following section provides basic notation and formalization for the remainder of the paper. Section 3 then introduces the rationale, design, and estimation method of the new precision measure. Subsequently, Sect. 4 provides the empirical evaluation and discussion. Finally, after a brief overview of the related work, the overall conclusions are provided in Sect. 6

2   Preliminaries

Activities and Traces. A process consists of activities that are represented by their activity labels. The set of all possible activity labels in the system form the activity alphabet $\mathscr{A}$. A trace $\sigma_i \in \mathscr{A}^*$ is a sequence of activity labels, where $\mathscr{A}^*$ is the set of all finite sequences over $\mathscr{A}$. The length of a trace is represented as $|\sigma_i|$.

System. The system $\mathcal{S} = (S, \pi_S)$ represents the underlying process and consists of two components. The first component is the support of the system $S \subseteq \mathscr{A}^*$ which consists of all traces that can be produced by the system. The second component is the system probability distribution $\pi_S : S \rightarrow (0, 1]$, which is a categorical distribution and assigns a probability of occurrence $\pi_S(\sigma_i)$ to each trace σ_i in the system support S, such that $\sum_{\sigma_i \in S} \pi_S(\sigma_i) = 1$. The size of the system corresponds to the number of traces σ_i in the system support and is represented as $K = |S|$. Note that we thus assume S to be finite.

Log. The log $L \in \mathcal{B}(\mathscr{A}^*)$ is a multi-set of traces. The frequency of trace σ_i in the log is denoted by n_{σ_i}, and $N = \sum_{\sigma_i \in L} n_{\sigma_i}$ denotes the size of the log. Note that, as we interpret the system as a categorical distribution over the system support S, the log L is a sample of N drawings from this distribution π_S.

Model. Two type of process models are considered: non-probabilistic and stochastic process models. A non-probabilistic model does not hold any information about the model probability $\pi_M(\sigma_i)$ of a trace σ_i and is simply a set of traces $M \subseteq \mathscr{A}^*$. A stochastic process model $\mathcal{M} = (M, \pi_M)$ is more informative as it consists of a set of traces $M \subseteq \mathscr{A}^*$ that represents the models support and a model probability distribution $\pi_M : M \rightarrow (0, 1]$ where $\pi_M(\sigma_i)$ represents the probability of trace σ_i according to the model, such that $\sum_{\sigma_i \in M} \pi_M(\sigma_i) = 1$. The number of traces in the model is denoted as $|M|$.

3   Alpha Precision

The overall goal is to design a model-system precision measure for the use case of discovering a process model which only contains significant system behavior. Furthermore, the measure should meet the following three design requirements:

Requirement 1. *The precision measure should quantify to which extent the process model (only) contains significant process behavior.*

Requirement 2. *The precision measure should quantify the precision of the model with respect to the system.*

Requirement 3. *The precision measure should produce values that have an unambiguous and human-interpretable meaning.*

3.1 Rationale and Design

In order to meet the first two design requirements, the following definition first introduces the concept of α-significance, which identifies a trace as significant if its system probability $\pi_s(\sigma_i)$ exceeds a user-defined threshold α.

Definition 1 (α-significance). *A trace σ_i is α-significant iff $\pi_s(\sigma_i) > \alpha$*

Based on this concept, we can define the α-indicator function I_α.

Definition 2 (α-indicator function).

$$I_\alpha(\sigma_i) = \begin{cases} 1, & \text{if } \pi_S(\sigma_i) \geq \alpha \\ 0, & \text{otherwise} \end{cases} \quad (1)$$

For a stochastic model, $\mathcal{M}$, α-precision P_α is then defined as the probability that the model produces a trace that is α-significant.

Definition 3 (α-precision (stochastic model)). *Let $\mathcal{M}$ be a stochastic process model, then*

$$P_\alpha(\mathcal{M}, \mathcal{S}) = \sum_{\sigma_i \in M} \pi_M(\sigma_i) I_\alpha(\sigma_i) \quad (2)$$

For non-probabilistic models, α-precision is defined as the portion of α-significant traces in the model M.

Definition 4 (α-precision (non-probabilistic model)). *Let M be a non-probabilistic process model, then*

$$P_\alpha(M, \mathcal{S}) = \frac{1}{|M|} \sum_{\sigma_i \in M} I_\alpha(\sigma_i) \quad (3)$$

The third design requirement involves a meaningful and clear interpretation of the new precision measure. In order to illustrate the interpretability of the proposed α-precision, consider the following application scenario:

A data scientist wants to discover a process model from an event log that contains the significant (typical) behavior. The goal is to understand the standard way of working within the department and not depict exceptional process executions in the process model. First, they set the α threshold at 1%, which means that any trace that has a probability less than 1% is considered non-significant. Next, the discovered stochastic model appears to have an $P_\alpha = 0.8$. This value tells her that 80% of the traces generated by this process model are expected to be significant, i.e., 80% of the behavior produced by the model has a system probability $\pi_S(\sigma_i)$ greater than 1%.

Note how both the α-threshold and the α-precision have natural interpretations that allow users to use context and domain-expertise to set a proper threshold and interpret and evaluate precision levels found for their discovered models. Also, note that in the case of a deterministic model, the interpretation of the precision measure would only slightly change to the conclusion that 80% of the traces contained in the model are significant.

3.2   Estimation Method

Calculating the α-precision is straightforward when the system $\mathcal{S}$ and the its probability distribution π_S are known. However, in real-life, one does not know the system. The only available information is typically an event log—a limited sample of the system's behavior. Therefore, we introduce a method to estimate the true α-precision from the available information in the log.

The general idea behind the method is to estimate the system probabilities $\widehat{\pi_S}$ from the event log. Next, these estimates are used to estimate the indicator function (cfr. Eq. 1), which subsequently is used to estimate the α-precision $\widehat{P_\alpha}$. Thus, the estimation problem reduces to the estimation of the system probabilities from the event log. The proposed method is based on Bayesian Inference and inspired by the work of [10].

To infer knowledge about the system from the log, we need additional assumptions about the system.

Assumption 1. *The system contains a finite amount of behavior.*

This assumption implies that the system has some mechanism that prevents a process from being executed indefinitely. For business processes with humans involved, this is a fair assumption.

Assumption 2. *The system support is correctly defined.*

This assumption restricts the modeling of uncertainty to the system probability density. The assumption that the system support is correctly defined corresponds to the common assumption in statistics of correct model specification. While it is hard to prove that the system support is defined correctly, it is essential to realize that any theoretically possible trace is part of the system support, no matter how small the probability of occurrence.

Assumption 3. *The log is a representative sample from the system's behavior*

As the proposed method will rely on statistical inference, this assumption is required to draw proper conclusions from the data for the underlying system. Considering these assumptions, the α-precision can be estimated in four steps.

Step 1: Define the System Support. First, the system support S of the system $\mathcal{S}$ needs to be specified. Under the assumption that the system behavior is finite, the system support can be defined as a set of traces σ_i for which the system probability $\pi_S(\sigma_i) > 0$.

Step 2: Define a Prior Distribution over $\mathcal{S}$. As indicated before, a system $\mathcal{S} = (S, \pi_S)$ consists of two components: its support and a probability function. The latter can be defined as the vector $\pi_S = (\pi_S(\sigma_1), \ldots, \pi_S(\sigma_K))$ of system probabilities, where K is the size of the system.

However, because the actual system is unknown, we do not know the true categorical probability function. In order to model this uncertainty, we consider all theoretically possible categorical probability functions for K possible outcomes and assign a probability to each one of them. This is modeled as a Dirichlet distribution [12].

From the perspective of Bayesian Inference, the first step is to encode the prior belief about the system as the prior distribution. In this paper's context, the prior belief refers to the knowledge about the system probability function π_S before observing the data. Assuming that there is no specific information to favor one probability function over the other, a flat Dirichlet distribution is chosen as the prior. This distribution is equivalent to a uniform distribution over all possible system probability distributions π_S and achieved by setting all the Dirichlet parameters θ_i to 1.

Step 3: Determine Posterior Distributions. The flat Dirichlet distribution from the previous step represents our prior belief that all possible probability functions π_S are equally likely. However, once we have observed an event log, we notice that some traces are more common than others, indicating that some probability functions π_S must be more likely than others.

Bayesian inference uses Bayes' theorem to update our prior beliefs with the evidence in the log, which results in a posterior distribution [6]. In most situations, the posterior distribution is not obtainable analytically unless the prior distribution is conjugate to the likelihood distribution of the data. If this is the case, the posterior distribution can be analytically calculated from the prior distribution and the data.

Because the Dirichlet distribution is a conjugate prior to the multinomial distribution and the event log is a multinomial distribution, the posterior distribution is also a Dirichlet distribution. More specifically, the posterior distribution will be a Dirichlet distribution with parameters $\theta'_i = \theta_i + n_i$, where θ_i represents the i-th parameter of the prior distribution and n_i represents how often outcome i was observed in the data.

Given a flat Dirichlet distribution as the prior distribution and our event log L being a multi-set of traces σ_i, this results in the following posterior distribution:

$$\pi_S | L \sim \text{Dir}\left((1 + n_{\sigma_1}, \ldots, 1 + n_{\sigma_K})\right) \tag{4}$$

Note that the posterior distribution assigns a probability to each possible system probability distribution π_S based on the evidence in the log.

Step 4: Estimate α-Precision. Now that the posterior distribution over all possible system probability functions is known, the true system probability function can be estimated by taking the expected value of the Dirichlet posterior, which is defined as follows:

Definition 5. *Let $X = (X_1, \ldots X_K) \sim \text{Dir}(\theta')$, then the expected value of X_i is*

$$E[X_i] = \frac{\theta'_i}{\sum_{k=1}^{K} \theta_k} \tag{5}$$

Given that the parameters of the posterior distribution are $\theta'_i = 1 + n_{\sigma_i}$ for $1 \leq i \leq K$, we can estimate the system probabilities $\pi_S(\sigma_i)$ as follows:

$$\forall \sigma_i \in S : \widehat{\pi_S(\sigma_i)} = \frac{1 + n_{\sigma_i}}{\sum_{k=1}^{K}(1 + n_{\sigma_k})} = \frac{1 + n_{\sigma_i}}{K + N} \tag{6}$$

Based on this estimator for system probability, we can subsequently estimate the α-significance indicator function as follows:

$$\widehat{I_\alpha(\sigma_i)} = \begin{cases} 1, & \text{if } \widehat{\pi_S(\sigma_i)} \geq \alpha \\ 0, & \text{otherwise} \end{cases} \tag{7}$$

Using this indicator function in Eqs. 2 and 3, for stochastic and non-probabilistic models respectively, we can then estimate the α-precision. We will denote this estimated precision as $\widehat{P_\alpha}$.

3.3 Estimating the System Support Size

Analyzing the final equations of the proposed estimation method reveals that it requires two parameters: the α-significance level and the theoretical system support size K. While the former can be freely chosen and should reflect the domain expert's interpretation of significance, the latter should preferably match its theoretical value. Various approaches to estimate K can be devised. This paper proposes two approaches—the *unrestricted* and *restricted* approach—and motivates them both based on underlying assumptions.

The unrestricted approach is so-called because it does not strongly limit the behavior that is included in the system support, except for the alphabet of activity labels observed in the event log $\mathscr{A}$ and a user-defined maximum trace length γ. It then defines the system support S as the set of all possible sequences σ_i over $\mathscr{A}$ with a length $|\sigma_i| \leq \gamma$. This approach assumes that the entire activity alphabet has been observed and a maximum trace length exists. The rationale behind the maximum trace length assumption is that the system would not allow a process instance to keep ongoing indefinitely. Given the alphabet $\mathscr{A}$ and the maximum trace length γ, the size of the system can be calculated as follows: $K = \sum_{i=1}^{\gamma} \mathscr{A}^i$.

The restricted approach can be seen as taking the system support S from the first approach as its starting point but removing all traces that contain a directly-follows relation not observed in the log. The assumption thus is that all possible directly-follows relations have been observed in the log.

Suppose that the directly-follows relations are represented by a matrix $\mathbf{D}$ of size $|\mathscr{A}| \times |\mathscr{A}|$ where D_{ij} equals 1 if and only if it was observed in the log that activity i of the alphabet was directly followed by the activity j, and 0 otherwise.

Table 1. Parameters.

	Parameter	Values
System	Alphabet length	$[4, 6, 8]$
	Max trace length	$[4, 6]$
Log	Log size	$[100, 500, 1000, 5000, 10000]$
Model	Discovery threshold	$[0.3, 0.4, 0.5, 0.6, 0.7]$
	# iterations	25

It follows naturally that the number of allowed sequences of length two is equal to the sum over this matrix, i.e., $\sum_{\mathbf{D}} D_{ij}$. Furthermore, assume a vector $\mathbf{o}$ of size $|\mathscr{A}|$, where o_i is 1 if and only if activity i in the alphabet is a valid start activity, and 0 otherwise. The number of sequences of length one with a valid start activity is then equal to the sum over vector $\mathbf{o}$, i.e., $\sum_{\mathbf{o}} o_i$. The number of sequences of length two with a valid start activity is equal to the sum over $\mathbf{o}^{\mathsf{T}} \mathbf{D}$. This can be generalized to $\mathbf{o}^{\mathsf{T}} \mathbf{D}^{\gamma-1}$ for sequences of length γ. In order to limit the number of sequences to specific final activities, a vector $\mathbf{f}$ of size $|\mathscr{A}|$ can be defined where f_i is 1 if and only if activity i is a valid end activity, and 0 otherwise. The scalar $\mathbf{o}^{\mathsf{T}} \mathbf{D}^{\gamma-1} \mathbf{f}$ then equals the total number of sequences of length γ with valid start and end points.[1] As a result, for the restricted approach, $K = \sum_{i=1}^{\gamma} \mathbf{o}^{\mathsf{T}} \mathbf{D}^{i-1} \mathbf{f}$, which is computationally easy to calculate. Note that the restriction of valid start and end activities can be omitted without difficulty depending on the specific context.

4   Empirical Evaluation

This Section provides an empirical evaluation of the α-precision by means of a controlled experiment and a application on real-life data. The goal is to provide insights and knowledge claims about the behavior and performance of this newly developed measure.

4.1   Unbiasedness Estimator

In this section, we describe a controlled experiment to analyze the unbiasedness of the α-precision estimator. The experiment exists of the following six steps: (1) Generate systems, (2) Generate logs, (3) Generate models, (4) Calculate actual α-precision, (5) Calculate estimated α-precision, and (6) Analyze bias.

Generate Systems. First, different systems were generated using the alphabet length and maximum trace lengths in Table 1 as input parameters. Each of the system-traces is assigned a probability $\pi_S(\sigma_i)$.

---

[1] In the specific case that $\gamma = 1$, $\mathbf{D}^0$ equals the identity matrix $\mathbf{I}$, and thus $\mathbf{o}^{\mathsf{T}} \mathbf{D}^{\gamma-1} \mathbf{f} = \mathbf{o}^{\mathsf{T}} \mathbf{f}$, which is the number of activities that are both valid start and end activities. This is indeed equal to the number of valid sequences of length one.

Because we want define the full system including probability distribution in this supervised experiment, both the size of the alphabet and the maximum possible trace length are kept relatively low. An alphabet of length 10 with a maximum trace length of 8 leads to 1.23×10^9 possible systems sequences.

Generate Logs. For each of the systems, logs of different sizes (cfr. Table 1) are generated, using the defined system probabilities for each trace.

Generate Models. Subsequently, a model is generated based on each system. A discovery threshold is set to steer the amount of the system that is captured by the model. A discovery threshold of 0.7 means that each trace has a 70% probability to be included in the model.

To each of the traces that is included in the model, a probability is assigned in order to turn it into a stochastic model as defined in Sect. 2. These probabilities are independent from the system probabilities created in step 1, thereby making sure no algorithm bias is introduced.

For each of the combinations of parameters listed in Table 1, we repeat this process 25 times.

Calculate Actual α-Precision. Given that we know the actual system probability distribution π_S, the actual α-precision can be computed using Eq. 2. Because there is no domain expertise in this artificial setting to define the level of α, a rule of thumb was used to set α equal to $\frac{1}{K}$ where K is the size of the system support.

Calculate Estimated α-Precision. Given the model and log, we can then estimate the α-precision by using Eq. 2 in combination with the estimated system probabilities as defined in Eq. 6.

Next to the information provided in the model and the log, we need to define the value of K and α. For K, the size of the system, we take the *unrestricted* approach described in Step 1 of Sect. 3.2, where we consider all possible sequence of the alphabet. Subsequently, α is also set to $\frac{1}{K}$ for the estimation.

Measure Bias. Given both the actual α-precision and the estimated α-precision, we define the difference between the two as follows.

$$\beta = \widehat{P_\alpha} - P_\alpha \tag{8}$$

When the estimated α-precision is greater than the actual α-precision, the bias as measured by β will be positive, and there is thus an overestimation. Otherwise, β will be negative and the actual α-precision will thus be underestimated.

Results. In Fig. 1, it can be seen that the estimator underestimates the real precision when the size of the log is relatively small compared to the size of the system. For the system with size 340 (i.e. alphabet length of 4 and maximum trace length of 4), the estimator becomes unbiased when approximately

1000 cases have been observed in the log. For larger systems, the biases only approaches zero for logs of 10000 cases, while for the largest systems in the experiment the estimators still shows a large bias at logs of size 10000.

Figure 2 shows the extent of bias specifically in relation to the ratio between the log size and the system size. The vertical line indicates the where the ratio is 1, i.e. the number of different sequences in the system support equals the number of observed traces in the log. It can be seen that the biases quickly decreases when the ratio approaches 1, and then decreases more gradually toward zero for ratios greater than 1.

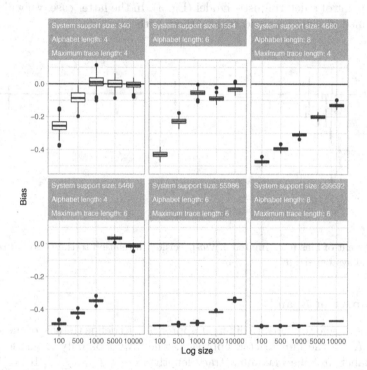

Fig. 1. Bias $\widehat{P_\alpha} - P_\alpha$ for different sizes of the log and the system.

Table 2. BPIC'12 log - descriptive statistics.

Metric	Value	Metric	Value
Number of activities	10	Number of traces	17
Number of events	60849	Max. trace length	8
Number of cases	13087	Avg. trace length	4.65

4.2 Demonstration on Real-Life Event Logs

In this and the next section, the proposed method will be applied on real-life
event data. For this, data from the Business Process Intelligence Challenge 2012
is used [5]. Descriptive statistics for this event log can be found in Table 2.

Based on the log, a stochastic model has been discovered using the frequency
estimator [4] on a model discovered by the Directly Follows Model Miner [18].
The discovered model contains 6 different activity sequences, of which the probability varies between 0.029 and 0.509. In the analyses, we will both approach
the estimation from the starting point of a stochastic model (Eq. 2), as from the
starting point of a deterministic model (Eg. 3). In the latter case, we will ignore
the obtained probabilities and replace them with $\frac{1}{|M|}$.

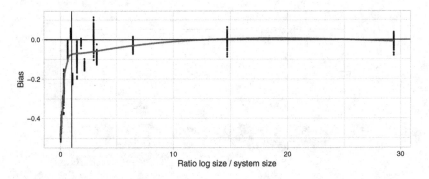

Fig. 2. Extent of bias in estimated *alpha*-precision in relation to the ratio between log
size N and system size K.

4.3 Impact of K and α

As discussed before, there are different approaches to define the size of the system
support K. In the unrestricted approach, where we look only at the length of
the alphabet and the maximum trace length, we get $K = \sum_{i=1}^{8} 10^i = 1.11 \times$
10^8. Given the fairly high structuredness of the data in question, this seems an
exuberant amount. If we therefore take the restricted approach instead, where
we only take into account sequences that adhere to the observed directly-follows
relations, start activities and end activities, the system support K' is only 32.
When we drop the start and end activity requirement, K'' equals 1539.

Figure 3 shows the value of $\widehat{P_\alpha}$ for different values of K and different values
of α for both the stochastic and deterministic approach. It can be seen that the
estimated α-precision is relatively stable with respect to the value of K. Only
when K is increased to 50000 can noticeable differences in $\widehat{P_\alpha}$ be seen.

Note that as the model only contains 6 different activity sequences, we can
see apparent jumps in the measured precision when α changes such that a traces
moved from insignificant to significant. For the deterministic model all jumps
are equal in size, while this is naturally not the case for the stochastic model.

The biggest jump in the latter case happens when α drops below 0.05 (approximately), and the trace with the highest probability according to the model (0.509) becomes significant.

Impact of Significance Filtering. Figure 4 shows the values of $\widehat{P_\alpha}$ for models discovery by the Inductive Miner infrequent [16] and Directly-follows miner [18], with different setting for significance filtering. The higher the significance parameter, the more significant behavior must be to make it into the model.

It can be seen that, for different values of alpha, when the significance parameters increases, so does the estimated precision. This provides implicit validation that the proposed measure behaves as expected, as models with a stronger significance filtering gets higher precision scores.

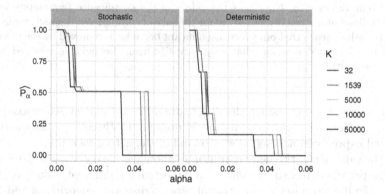

Fig. 3. $\widehat{P_\alpha}$ for different values of K and alpha.

Figure 4 also shows the potential of the α-precision to analyze and compare various algorithms. The visual analysis shows that for the DFM-algorithm, the evolution of precision with respect to the significance parameter is smoother than for the Inductive Miner algorithm. This implies better control for significance filtering in the former algorithm. This derives from the fact that Inductive miner often results in the same model for various significance filtering levels. In the extreme, the measured precision drops to zero for the DFM algorithm with the filtering parameter set to 100, as this results in an empty model.

4.4   Discussion

Based on the design and empirical evaluation of the α-precision measure and its estimation method, various knowledge insights can be constructed about the precision measure.

The controlled experiment showed that the measure is unbiased when the size of the log is sufficiently large in comparison with the system sizes. When

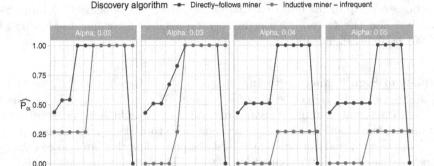

Fig. 4. $\widehat{P_\alpha}$ for different values in the significance filtering parameter of [16] and [18], and different values of α. Note that the values of the significance parameters for the Directly-follows miner has been inverted for the sake of comparison. A small significance parameter value means that only very insignificant behavior is removed by the discovery algorithm, a large value means that only very significant behavior is retained by the discovery algorithms.

there are fewer cases in the log than there are different activity sequences in the system, a substantial underestimation is present. In those circumstances, the estimated α-precision acts as a lower bound of the actual measure.

At the same time, the demonstration of the measure on real-life data shows that the proposed measures behave as expected for different values of K, α and the significance filter parameters of process discovery algorithms and holds potential to evaluate process discovery algorithms aimed at discovering significant behavior. While defining the system support size K is an important step towards estimating the α-precision, it has been shown that the estimator is relatively robust for changes in K.

An ongoing discussion in the field of process mining is that of (desirable) properties (or axioms) of conformance measures [21]. However, the properties studied in past research are not applicable to α-precision as most properties are defined with respect to non-stochastic process models. The exception are the eight properties for stochastic conformance checking defined in [17]. Unfortunately, even these properties are not directly applicable because they relate to log-model measures, while α-precision is a system-model measure. Nevertheless, we can derive four properties for the α-precision which are inspired by the properties discussed in [17].

Property 1. The α-precision measure is deterministic.

This property relates directly to property P1 in [17] and holds as α-precision is a function of the assumed system support size K, the α-significance level, the log size N and the trace frequencies n_{σ_i}, which are all fixed at the start.

Property 2. The α-precision measure depends only on the stochastic language of the log and model and not on their representation.

This property refers directly to property P2 in [17] and holds naturally, as the α-precision is calculated directly from the stochastic language of the log and model.

Property 3. The α-precision measure returns values between 0 and 1.

This property relates directly to property P3 in [17]. Since the α-indicator function $\widehat{I_\alpha}(x)$ is either 0 or 1 (cf. Eq. 7), it follows that the minimum and maximum value of the α-precision (cf. Definitions 3 and 4) is also 0 and 1 and can only be achieved when all α-indicator functions evaluate to 0 or 1 respectively.

Property 4. The α-precision measure asymptotically goes to 1 if (i) the model only contains the α-significant system behavior, (ii) the log has the same stochastic language as the system and (iii) the log size increases towards infinity.

This property is an adaptation of property P4 in [17] to the context of our measure. If the log size N goes towards infinity, then the system probability estimates (cf. Eq. 6) will go towards $\frac{n_{\sigma_i}}{N}$. The latter equates to the true system probabilities since the log and the system express the same stochastic language. Consequently, if the model only contains α-significant system behavior, the indicator function will evaluate to 1 for all traces in the model and the α-precision measure will equate to 1.

The other properties P5, P7 and P8 in [17] are not directly applicable to our measure. Properties P5, P6 and P8 are related to recall measures rather than precision measures. While property P6 does relate to precision measures, it focuses on a log-model relation which doesn't have a clear analog counterpart in the system-model context we are operating.

5   Related Work

In typical process mining projects, the system is unknown; thus, quality measures (conformance checking techniques) have focused on the relationship between model and log, rather than system and model. The quality of (non-stochastic) models with respect to logs is typically measured using fitness, precision, generalization, and simplicity, where fitness is the fraction of behavior of the log that is in the system, precision is the fraction of behavior in the model that was observed in the log, generalization is the predicted fraction of future behavior of the system that is in the model, and simplicity expresses the size or complexity of the model to express its behavior [3]. Recently, the concept of precision (and to a lesser extent generalization) has seen discussion in terms of desirable properties such measures should possess [21]; however, this discussion has not yet included unbiasedness with respect to unknown systems. Of these quality dimensions, generalization aims to describe the system and could be seen as a system-fitness measure [20]. Such log-based measures are not unbiased estimators of system properties empirically [9]. Compared to these approaches, our proposed measure explicitly and understandably takes the significant behavior of an unknown system into account.

For stochastic process models, quality measures include stochastic distance [13], stochastic precision and recall [17], and entropic relevance [19], however these do not aim to compare a model with an unknown system. While not intended for the system-model context, it would be interesting to study the bias of these techniques when applied in a system-model context, like [9,11].

The system has been the subject of study in process mining, as the ultimate goal of process mining is to obtain insights into the system to improve it. Some process discovery techniques guarantee to return a model that is the language equivalent to the system, under some assumptions, such as the log being noise-free or complete with respect to a particular abstraction of the system [1,14,23]. However, such techniques do not offer any guarantees when these assumptions are not met; thus, it is a valuable exercise to have an unbiased estimator of the relation between system and model.

6   Conclusions

In process mining, organizations aim to gain insights into their business processes, which we refer to as systems, by discovering process models from event logs. Typically, the quality of a process model is assessed with respect to an event log, however we argue that it might be useful to compare a model to the unknown system, based on its significant behavior. In this paper, we presented a new precision metric that expresses the extent to which the model contains significant system behavior, based on an α-significance level. We empirically evaluated the new measure by showing that it can be unbiased under certain assumptions and demonstrated its applicability and value on real-life event logs.

While the initial results indicate that this precision measure supports the analysis of discovery algorithms aimed at discovering significant behavior, the empirical analysis also shows that the construction of unbiased system estimators is particularly challenging and requires future research to better understand and remove this apparent bias. Important aspects to consider are a more realistic definition of ground-truth systems in controlled experiments, as well as the proper estimation of the system size when using the estimator. The impact of prior configurations, which are currently uninformative, is another aspects that requires further analyses.

Overall, we hope this work provides an initial yet solid foundation for further research into system measures supporting the use case of discovering significant behavior.

References

1. Badouel, E.: On the α-reconstructibility of workflow nets. In: Haddad, S., Pomello, L. (eds.) PETRI NETS 2012. LNCS, vol. 7347, pp. 128–147. Springer, Heidelberg (2012). https://doi.org/10.1007/978-3-642-31131-4_8

2. Brons, D., Scheepens, R., Fahland, D.: Striking a new balance in accuracy and simplicity with the probabilistic inductive miner. In: ICPM, pp. 32–39. IEEE (2021)
3. Buijs, J.C.A.M., van Dongen, B.F., van der Aalst, W.M.P.: Quality dimensions in process discovery: the importance of fitness, precision, generalization and simplicity. Int. J. Cooper. Inf. Syst. **23**(1) (2014)
4. Burke, A., Leemans, S.J.J., Wynn, M.T.: Stochastic process discovery by weight estimation. In: Leemans, S., Leopold, H. (eds.) ICPM 2020. LNBIP, vol. 406, pp. 260–272. Springer, Cham (2021). https://doi.org/10.1007/978-3-030-72693-5_20
5. van Dongen, B.: BPI challenge 2012, April 2012. https://doi.org/10.4121/uuid: 3926db30-f712-4394-aebc-75976070e91f
6. Gelman, A., Carlin, J.B., Stern, H.S., Dunson, D.B., Vehtari, A., Rubin, D.B.: Bayesian Data Analysis. Taylor & Francis Ltd, Boca Raton (2013)
7. Günther, C.W., van der Aalst, W.M.P.: Fuzzy mining – adaptive process simplification based on multi-perspective metrics. In: Alonso, G., Dadam, P., Rosemann, M. (eds.) BPM 2007. LNCS, vol. 4714, pp. 328–343. Springer, Heidelberg (2007). https://doi.org/10.1007/978-3-540-75183-0_24
8. Günther, C.W., Rozinat, A.: Disco: discover your processes. In: BPM. CEUR Workshop Proceedings, vol. 940, pp. 40–44. CEUR-WS.org (2012)
9. Janssenswillen, G., Depaire, B.: Towards confirmatory process discovery: making assertions about the underlying system. Bus. Inf. Syst. Eng. **61**(6), 713–728 (2019)
10. Janssenswillen, G., Depaire, B., Faes, C.: Enhancing discovered process models using Bayesian inference and MCMC. In: Del Río Ortega, A., Leopold, H., Santoro, F.M. (eds.) BPM 2020. LNBIP, vol. 397, pp. 295–307. Springer, Cham (2020). https://doi.org/10.1007/978-3-030-66498-5_22
11. Janssenswillen, G., Jouck, T., Creemers, M., Depaire, B.: Measuring the quality of models with respect to the underlying system: an empirical study. In: La Rosa, M., Loos, P., Pastor, O. (eds.) BPM 2016. LNCS, vol. 9850, pp. 73–89. Springer, Cham (2016). https://doi.org/10.1007/978-3-319-45348-4_5
12. Kotz, S., Balakrishnan, N., Johnson, N.L.: Continuous Multivariate Distributions: Models and Applications, vol. 1. Wiley, Hoboken (2004)
13. Leemans, S.J.J., van der Aalst, W.M.P., Brockhoff, T., Polyvyanyy, A.: Stochastic process mining: earth movers' stochastic conformance. Inf. Syst. **102**, 101724 (2021)
14. Leemans, S.J.J., Fahland, D.: Information-preserving abstractions of event data in process mining. Knowl. Inf. Syst. **62**(3), 1143–1197 (2019). https://doi.org/10. 1007/s10115-019-01376-9
15. Leemans, S.J.J., Fahland, D., van der Aalst, W.M.P.: Discovering block-structured process models from event logs containing infrequent behaviour. In: Lohmann, N., Song, M., Wohed, P. (eds.) BPM 2013. LNBIP, vol. 171, pp. 66–78. Springer, Cham (2014). https://doi.org/10.1007/978-3-319-06257-0_6
16. Leemans, S.J.J., Fahland, D., van der Aalst, W.M.P.: Discovering block-structured process models from incomplete event logs. In: Ciardo, G., Kindler, E. (eds.) PETRI NETS 2014. LNCS, vol. 8489, pp. 91–110. Springer, Cham (2014). https:// doi.org/10.1007/978-3-319-07734-5_6
17. Leemans, S.J.J., Polyvyanyy, A.: Stochastic-aware conformance checking: an entropy-based approach. In: Dustdar, S., Yu, E., Salinesi, C., Rieu, D., Pant, V. (eds.) CAiSE 2020. LNCS, vol. 12127, pp. 217–233. Springer, Cham (2020). https://doi.org/10.1007/978-3-030-49435-3_14
18. Leemans, S.J.J., Poppe, E., Wynn, M.T.: Directly follows-based process mining: exploration & a case study. In: ICPM, pp. 25–32. IEEE (2019)

19. Polyvyanyy, A., Moffat, A., García-Bañuelos, L.: An entropic relevance measure for stochastic conformance checking in process mining. In: ICPM, pp. 97–104. IEEE (2020)

20. Syring, A.F., Tax, N., van der Aalst, W.M.P.: Evaluating conformance measures in process mining using conformance propositions. Trans. Petri Nets Other Model. Concurr. **14**, 192–221 (2019)

21. Tax, N., Lu, X., Sidorova, N., Fahland, D., van der Aalst, W.M.P.: The imprecisions of precision measures in process mining. Inf. Process. Lett. **135**, 1–8 (2018)

22. Weijters, A., van Der Aalst, W.M., De Medeiros, A.A.: Process mining with the heuristics miner-algorithm. Technical report WP 166, Technische Universiteit Eindhoven, pp. 1–34 (2006)

23. van Zelst, S.J., van Dongen, B.F., van der Aalst, W.M.P.: Avoiding over-fitting in ILP-based process discovery. In: Motahari-Nezhad, H.R., Recker, J., Weidlich, M. (eds.) BPM 2015. LNCS, vol. 9253, pp. 163–171. Springer, Cham (2015). https://doi.org/10.1007/978-3-319-23063-4_10

Learning to Act: A Reinforcement Learning Approach to Recommend the Best Next Activities

Stefano Branchi[1], Chiara Di Francescomarino[1], Chiara Ghidini[1], David Massimo[2], Francesco Ricci[2], and Massimiliano Ronzani[1(✉)]

[1] Fondazione Bruno Kessler, Trento, Italy
{sbranchi,dfmchiara,ghidini,mronzani}@fbk.eu
[2] Free University of Bozen-Bolzano, Bolzano, Italy
{David.Massimo,fricci}@unibz.it

Abstract. The rise of process data availability has recently led to the development of data-driven learning approaches. However, most of these approaches restrict the use of the learned model to predict the future of ongoing process executions. The goal of this paper is moving a step forward and leveraging available data to *learning to act*, by supporting users with recommendations derived from an optimal strategy (measure of performance). We take the optimization perspective of one process actor and we recommend the best activities to execute next, in response to what happens in a complex external environment, where there is no control on exogenous factors. To this aim, we investigate an approach that learns, by means of Reinforcement Learning, the optimal policy from the observation of past executions and recommends the best activities to carry on for optimizing a Key Performance Indicator of interest. The validity of the approach is demonstrated on two scenarios taken from real-life data.

Keywords: Prescriptive Process Monitoring · Reinforcement Learning · Next activity recommendations

1 Introduction

In the last few years, a number of works have proposed approaches, solutions and benchmarks in the field of Predictive Process Monitoring [3,14]. Predictive Process Monitoring leverages the analysis of historical execution traces in order to predict the unrolling of a process instance that has been only partially executed. However, most of these efforts have not used the predictions to explicitly support user with recommendations, i.e., with a concrete usage of these predictions. In fact, there is a clear need of *actionable* process management systems [7] able to support the users with recommendations about the best actions to take.

The overall goal of this paper is therefore moving a step forward, towards the implementation of a *learning to act* system, in line with the ideas of Prescriptive

© Springer Nature Switzerland AG 2022
C. Di Ciccio et al. (Eds.): BPM 2022, LNBIP 458, pp. 137–154, 2022.
https://doi.org/10.1007/978-3-031-16171-1_9

Process Monitoring [8,17]. Given an ongoing business process execution, Prescriptive Process Monitoring aims at recommending activities or interventions with the goal of optimizing a target measure of interest or Key Performance Indicator (KPI). State-of-the-art works have introduced methods for raising alarms or triggering interventions, to prevent or mitigate undesired outcomes, as well as for recommending the best resource allocation. Only few of them have targeted the generation of recommendations of the next activity(ies) to optimize a certain KPI of interest [2,9,25], such as, the cycle time of the process execution. Moreover, none of them explicitly considers the process execution in the context of a complex environment that depends upon exogenous factors, including how the other process actors behave. In this setting, identifying the best strategy to follow for a target actor, is not straightforward.

In this paper, we take the perspective of one target actor and we propose a solution based on Reinforcement Learning (RL): to recommend to the actor what to do next in order to optimize a given KPI of interest for this actor. To this aim, we first learn, from past executions, the response of the environment (actions taken by other actors) to the target actor's actions, and we then leverage RL to recommend the best activities/actions to carry on to optimize the KPI.

In the remainder of the paper after introducing some background concepts (Sect. 2), we present two concrete Prescriptive Process Monitoring problems that we have targeted (Sect. 3). Section 4 shows how a Prescriptive Process Monitoring problem can be mapped into RL, while Sect. 5 applies the proposed RL approach to the considered problems and evaluates its effectiveness. Finally, Sect. 6 and Sect. 7 present related works and conclusions, respectively.

2   Background

2.1   Event Logs

An *event log* consists of traces representing executions of a process (a.k.a. a case). A trace is a sequence of *events*, each referring to the execution of an activity (a.k.a. an event class). Besides timestamps, indicating the time in which the event has occurred, events in a trace may have a data payload consisting of attributes, such as, the resource(s) involved in the execution of an activity, or other data recorded during the event. Some of these attributes do not change throughout the different events in the trace, i.e., they refer to the whole case (*trace attributes*); for instance, the personal data (*Birth date*) of a customer in a loan request process. Other attributes are specific of an event (*event attributes*), for instance, the employee who creates an offer (*resource*), which is specific of the activity `Create offer`.

2.2   Prescriptive Process Monitoring

Prescriptive Process Monitoring [8,17] is a branch of Process Mining that aims at suggesting activities or triggering interventions for a process execution for optimizing a desired Key Performance Indicator (KPI). Differently from

Predictive Process Monitoring approaches, which aim at predicting the future of an ongoing execution trace, Prescriptive Process Monitoring techniques aim at recommending the best interventions for achieving a target business goal. For instance, a bank could be interested in minimizing the cost of granting a loan to a customer. In such a scenario, the KPI of interest for the bank is the cost of the activities carried out by the bank's personnel in order to reach an agreement with the customer. The best actions that the bank should carry out to achieve the business goal (reaching the agreement while minimizing the processing time) can be recommended to the bank.

2.3   Reinforcement Learning

Reinforcement Learning (RL) [10,23] refers to techniques providing an intelligent agent the capability to act in an environment, while maximizing the total amount of reward received by its actions. At each time step t, the agent chooses and executes an *action* a in response to the observation of the *state* of the environment s. The action execution causes, at the next time step $t + 1$, the environment to stochastically move to a new state s', and gives the agent a *reward* $r_{t+1} = \mathcal{R}(s, a, s')$ that indicates how well the agent has performed. The probability that, given the current state s and the action a, the environment moves into the new state s' is given by the state transition function $\mathcal{P}(s, a, s')$. The learning problem is therefore described as a discrete-time Markov Decision Process (MDP), which is formally defined by a tuple $M = (\mathcal{S}, \mathcal{A}, \mathcal{P}, \mathcal{R}, \gamma)$:

- $\mathcal{S}$ is the set of states.
- $\mathcal{A}$ is the set of agent's actions.
- $\mathcal{P} : \mathcal{S} \times \mathcal{A} \times \mathcal{S} \to [0, 1]$ is the transition probability function. $\mathcal{P}(s, a, s') = Pr(s_{t+1} = s' | s_t = s, a_t = a)$ is the probability of transition (at time t) from state s to state s' under action $a \in \mathcal{A}$.
- $\mathcal{R} : \mathcal{S} \times \mathcal{A} \times \mathcal{S} \to \mathbb{R}$ is the reward function. $\mathcal{R}(s, a, s')$ is the immediate reward obtained by the transition from state s to s' with action a.
- $\gamma \in [0, 1]$ is a parameter that measures how much the future rewards are discounted with respect to the immediate reward. Values of γ lower than 1 model a decision maker that discount the reward obtained in the more distant future.[1]

An MDP satisfies the *Markov Property*, that is, given s_t and a_t, the next state s_{t+1} is conditionally independent from all prior states and actions and it only depends on the current state, i.e., $Pr(s_{t+1} | s_t, a_t) = Pr(s_{t+1} | s_0, \cdots, s_t, a_0, \cdots, a_t)$.

The goal of RL is computing a *policy* that allows the agent to maximize the cumulative reward. A policy $\pi : \mathcal{S} \to \mathcal{A}$ is a mapping from each state $s \in \mathcal{S}$ to an action $a \subset \mathcal{A}$, and the *cumulative reward* is the (discounted) sum of the rewards obtained by the agent while acting at the various time points t. The

---

[1] In this paper we set $\gamma = 1$, hence equally weighting the reward obtained at each action points of the target actor.

value of taking the action a in state s and then continuing to use the policy π, is the expected discounted cumulative reward of the agent, and it is given by the *state-action value function*: $Q^\pi(s, a) = \mathbb{E}_\pi(\Sigma_{k=0}^\infty \gamma^k r_{k+t+1} | s = s_t, a = a_t)$, where r_{t+1} is the reward obtained at time t. The *optimal* policy π^* dictates to a user in state s to perform the action that maximises $Q(s, \cdot)$. Hence, the optimal policy π^* maximises the cumulative reward that the user obtains by following the actions recommended by the policy π^*. Action-value functions can be estimated from experience, e.g., by averaging the actual returns for each state (action taken in that state), as with *Monte Carlo methods*.

Different algorithms can be used in RL [23]. Among them we can find the *value* and the *policy iteration* approaches. In the former the optimal action-value function $Q^*(s, a)$ is obtained by iteratively updating the estimate $Q^\pi(s, a)$. In the latter, the starting point is an arbitrary policy π that is iteratively evaluated (*evaluation phase*) and improved (*optimization phase*) until convergence. Monte Carlo methods are used in the policy evaluation phase for computing, given a policy π, for each state-action pair (s, a), the action-value function $Q^\pi(s, a)$. The estimate of the value of a given state-action pair (s, a) can be computed by averaging the sampled returns that originated from (s, a) over time. Given sufficient time, this procedure can construct a precise estimate Q of the action-value function Q^π. In the policy improvement step, the next policy is obtained by computing a greedy policy with respect to Q: given a state s, this new policy returns an action that maximizes $Q(s, \cdot)$.

3   Two Motivating Scenarios

We introduce here the considered problem by showcasing two real processes that involve one target actor, whose reward is to be maximised, and some more actors, contributing to determine the outcome of the process (environment).

Loan Request Handling (Loans). In a financial institute handling loan requests, customers send loan request applications and the bank decide either to decline an application, or to request further details to the customer, or to make an offer and start a negotiation with the customer. During the negotiation phase, the bank can contact the customer and possibly change its offer to encourage the customer to finally accept the bank's offer.

The bank aims at maximizing its payoff by trying to sign agreements with the customer, while reducing the costs of the negotiation phase, i.e., stopping negotiations that will not end up with an agreement. The bank is therefore interested to implement the best strategy to follow (actions) in order to maximize its interest.

Traffic fine management (Fines). In a police department in charge to collect road traffic fines, as in the scenario presented in [15], fines can be paid (partly or fully) either immediately after the fine has been issued, or after the fine notification is sent by the police to the offender's place of residence, or when the notification is received by the offender. If the entire amount is paid, the fine management process is closed. After being notified by post, the offender can appeal against the fine through a judge and/or the prefecture. If the appeal is successful, the case ends.

In such a setting, the police department aims at collecting the payment of the invoice by the offender as soon as possible, so as to avoid money wastes due to delays in payments or the involvement of the prefecture/judge. The department indeed receives credits for fast payments, no credits for payments never received and discredits for incorrect fines. The department is therefore interested to receive best action recommendations to maximize the received credits.

4   Mapping PPM to RL

We would like to support a target actor of interest in a process, such as, the financial institute or the police department (see Sect. 3), by providing them with recommendations for the best activities to execute in order to maximize their profit and their credits, respectively. To this aim, we leverage RL, by transforming the PPM problem of recommending the next activities to optimize a given KPI, into an RL problem, where the *agent* is the actor we are supporting in the decision making (e.g., the bank or the police department), and the *environment* is represented by the external factors—especially the activities carried out by the other actors involved in the process execution (e.g., the customer or the offender). We define our MDP so that:

- an *action*, to be recommended, is an activity of the actor of interest (*agent*) (e.g., the bank activity `Create offer`);
- a *state* is defined by taking into account the following variables:
  - the last activity executed by the actor of interest (e.g., the creation of a new offer by the bank) or by the other actors defining the stochastic response of the environment (e.g., the bank offer acceptance by the customer);
  - some relevant information about the history of the execution (e.g., the number of phone calls between the bank and the customer);
  - other aspects defining the stochastic response of the environment (e.g., the amount of the requested loan);
  A *state* is hence represented by a tuple $\langle LA, HF, EF \rangle$, where LA is the last activity executed by the actor of interest or by one of the other actors involved in the process, HF is a vector of features describing some relevant information of the process execution history and EF is a vector of features further describing the environment response to the actions of the actor of interest.
- the *reward function* is a numerical value that transforms the KPI of interest, computed on the complete execution, in a utility function at the level of single action.

Actions, states and reward function can be defined for each specific problem by leveraging the information contained in the event log and some domain

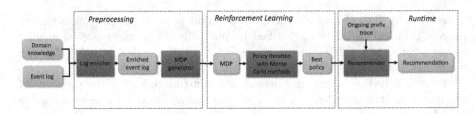

Fig. 1. Architecture of the RL solution

knowledge. The activities we are interested to recommend and those describing the stochastic response of the environment can be extracted from the event log. The relevant information about the history of the process execution can also be extracted from the event log, with some domain-specific pre-processing (e.g., counting the number of phone calls between the bank and the customer). The stochastic responses of the environment to the actor's actions can also be mined from the event log through trace attributes (e.g., the amount of the requested loan). Finally, information contained in event logs can be used to estimate the reward function for each state transition and action (e.g., in case the reward function is related to the process/event cycle time, the average duration of events of a certain type can be used to estimate the reward of a given state).

Figure 1 shows the architecture of the RL-based solution designed to solve the problem of recommending the next best activities to optimize a certain KPI. The input is an event log containing historical traces related to the execution of a process, and some domain knowledge, specifying the KPI of interest and the information that allows for the identification of actions, states and of the reward function. There are three main processing phases:

- *preprocessing phase:* the event log is preprocessed in order to learn a representation of the environment (i.e., the MDP). First, the event log is cleansed and the domain knowledge leveraged in order to annotate it. In detail, the event log is (i) filtered in order to remove low-frequency variants (with occurrence frequency lower than 10%) and activities that are not relevant for the decision making problem; (ii) enriched with attributes obtained by aggregating and preprocessing information related to the execution; (iii) annotated by specifying the agent's activities to be considered as actions; attributes and environment activities to be used for the state definition; attributes to be used for the computation of the reward function.

  Once the event log has been enriched and annotated, it can be used for building the MDP that defines the RL problem. To this aim, we start from the scenario-specific definition of action and state and, by replaying the traces in the event log, we build a directed graph, where each node corresponds to a state and each edge is labelled with the activity allowing to move from one node state to the other. Moreover, for each edge, the probability of reaching the target node (computed based on the number of traces in the event log that reach the corresponding state) and the value of the reward function are computed. Each edge is hence mapped to the tuple $(s, a, s', \mathcal{P}(s'|s, a), \mathcal{R})$ where s is the state corresponding to the source node of the edge, a is the action used for labelling the edge, s' is the state corresponding to the target node of the edge, $\mathcal{P}(s'|s, a)$ is computed as the percentage of the traces that reach the state s' among the traces that reach state s and execute a, and $\mathcal{R}$ is the value of the reward function.

- *RL phase:* the RL algorithm is actually applied to compute the optimal policy π^*; in this paper we used policy iteration with Monte Carlo methods.
- *runtime phase:* given an empty or ongoing execution trace, the policy is queried by the recommender system to return the best activities to be executed next.

5   Evaluation of the Recommendation Policy

We investigate the capability of the proposed approach to recommend the process activities that allow the target actor to maximize a KPI of interest, i.e., the optimal policy π^*, (i) when no activity has been executed yet, that is, the whole process execution is recommended; (ii) at different time steps of the process execution (i.e., at different prefix lengths), that is, when only a (remaining) part of the process execution is recommended. We hence explore the following research questions:

RQ1 How does the recommended sequence of activities (suggested by the optimal policy π^*) perform in terms of the KPI of interest when no activity has been executed yet?
RQ2 How does the recommended sequence of activities (suggested by the optimal policy π^*) perform in terms of the KPI of interest at a given point of the execution?

Unfortunately, the complexity of evaluating recommendations in the Prescriptive Process Monitoring domain is well known [6]. It relates to the difficulty to estimate the performance of recommendations that have possibly not been followed in practice. In order to answer our research questions, we therefore approximate the value of the KPI of interest (i) by leveraging a simulator (simulation evaluation); (ii) by looking at similar executions in the actual event log (test log evaluation). In the next subsections we describe the dataset (Sect. 5.1), we detail the experimental setting (Sect. 5.2), and we finally report the evaluation results (Sect. 5.3).

5.1   Datasets

We have used two real-world publicly-available datasets that, describing the behaviour of more than one actor, allow us to take the perspective of one of them (target): the BPI Challenge 2012 event log [4] (BPI2012) and the Road Traffic Fine Management event log [13] (FINES2015).

Table 1. Dataset description

Dataset	Trace #	Variant #	Event #	Event class #	Avg. trace length
BPI2012	13087	4366	262200	36	20
FINES2015	150370	231	561470	11	5

The BPI Challenge 2012 dataset relates to a Dutch Financial Institute. The process executions reported in the event log refer to an application process for personal loan (see the *Loans* scenario in Sect. 3). In this scenario we want to optimize the profit of the bank (*agent*), i.e., to minimize the cost C of granting a loan to a customer (*environment*) while maximizing the interest I of the bank granting the loan. To this aim, we define the KPI of interest for a given execution e as the difference between the amount of interest (if the bank offer is accepted and signed by the customer, namely if the activity `Offer accepted` occurs in the trace) and the cost of the employees working time, that is, the value of the KPI for the execution e is $\text{KPI}_{\text{BPI2012}}(e) = I(e) - C(e)$. The amount of interest depends on the amount class of the loan request: low (amount ≤ 6000), medium ($6000 <$ amount ≤ 15000) and high (amount > 15000). For the low class, the average interest rate is 16%, for the medium class, the average interest rate is 18%, while for the high class is 20%.[2] The cost of the employees' working time is computed assuming an average salary of 18 euros/h.[3]

The second dataset collects data related to an information system of the Italian police. The information system deals with the management of road traffic fines procedures, starting from the fine creation, up to the potential offender's appeal to the judge or to the prefecture (see the *Fines* scenario described in Sect. 3). Here, we want to maximize the credits received by the police department (*agent*) based on the fine payments received by the offender (*environment*). The department receives 3, 2 or 1 credits if the fines are fully paid within 6, within 12 months, or after 12 months respectively; it does not receive any credits if the fine is not fully paid, while it receives a discredit if the offender appeals to a judge or to the prefecture and wins, since these cases correspond to a money waste of the police authority. The KPI value for the execution e is $\text{KPI}_{\text{FINES2015}}(e)$, corresponding to the number of credits received for the execution.

Table 1 shows the number of traces, variants, events, event classes and average trace length of the considered datasets. Table 2 illustrates the MDP components for the two scenarios: the main MDP *actions*; the main MDP *state* components, i.e., the last activity (LA), the historical features (HF) and the environment features (EF); as well as the *reward*, including the main attributes used for its computation.[4]

---

[2] The information on the average interest rate is extracted from the BPI2017 [5] dataset which contains data from the same financial institution.

[3] We estimate the average salary of a bank employed in the Netherlands from https://www.salaryexpert.com/salary/job/banking-disbursement-clerk/netherlands.

[4] The complete MDP description is available at tinyurl.com/2p8aytrb.

For example, Table 3 shows how a trace related to the *Fines* scenario is pre-processed and transformed into an annotated trace, and then into MDP actions, states and rewards. The trace activities are annotated according to whether they have been carried out either by the *agent* or by the *environment*, and the attributes *2months* (the bimester since the fine creation), *amClass* (the fine amount class) and *payType* (type of payment performed) are computed. In the MDP construction step, the agent's activities (with the bimester interval[5]) are used as actions, while the state is built by leveraging the last executed activity (LA), the *2months* and the *amClass* attributes. The reward is not null when the payment is finally received and since in this trace the full payment is received after 6 months, 2 credits are awarded.

Once the log is enriched it is passed to the MDP generation step. We build two MDPs: the $MDP_{BPI2012}$ for the *Loan request handling* scenario (with 982 states and 15 actions) and the $MDP_{Fines2015}$ for the *Traffic fine management* scenario (with 215 sates and 70 actions).

5.2 Experimental Setting

In order to answer our research questions, the two event logs have been split in a training part, which is used in the *RL phase*, and a test part, which is used for the evaluation of the learned policy. For evaluating the computed policies, since in this setting both training and test set size can impact the evaluation results, we use two different splitting criteria (defining the percentage of event log used for the training and the test set): (i) 60%–40% (60% of the traces for the training set and 40% for the test set) and (ii) 80%–20% (80% for the training and 20% for the testing). For the evaluation of the optimal policy obtained by RL and for answering our two research questions, two different evaluations have been carried out: a simulation evaluation and a test log evaluation.

The simulation evaluation uses a Monte Carlo simulation similar to the one used in the training phase, but, differently from the training phase, where the MDP is obtained from the training log, here a test MDP, obtained from the test log, is leveraged to simulate the environment response. In this simulation, the optimal policy obtained from the RL approach is compared, in terms of the KPI of interest, against a random policy and against a policy corresponding to the most frequent decisions made by the actor in the actual traces. The value of the reward for each of the simulated policy is computed as the average over 100.000 simulated cases. This evaluation provides a preliminary answer to the first research question **RQ1**.

The test log evaluation aims at comparing the optimal policy obtained from RL with the actual policies used in the process. It is used for answering both our research questions. For **RQ1**, we focus on the policy recommended when no activity has been executed yet. In this setting, we compare the value of the

---

[5] The MDP actions in this scenario take into account, besides the activity name, also the 2-month interval (since the creation of the fine) in which the activity has been carried out (*2months*).

Table 2. MDP for the *Loan request handling* and the *Traffic fine management* scenarios.

Scenario	MDP description			
Loans	Action	Bank activities: loan acceptance, loan rejection, offer creation and delivery, requests for customer response		
	State	LA	Last activity of the agent (bank) or of the environment (customer)	
			Customer activities: application cancellation, offer sent back to the bank, offer acceptance	
		HF	call#	# of bank calls after the offer is sent
			miss#	# of requests for missing information
			offer#	# of offers to the customer
			reply#	# of customer replies to the offer
			fix	True if wrong inputs in the application are fixed
		EF	amClass	Loan amount class: low ($\leq$6000), medium and high (>15000)
	Reward	attr.	duration	Activity average duration
			amClass	Loan amount class
			granted	Whether the loan has been granted
		The reward is computed for each MDP state so that the reward of the complete execution corresponds to the value of the KPI$_{\text{BPI2012}}$ for that execution.[a]		
Fines	Action (See footnote 5)	Police department activities: fine creation and delivery, penalty increase and request for credit collection		
	State	LA	Last activity of the agent (police department) or of the environment (offender)	
			Offender activities: appeal to the Prefecture or to the Judge, payment	
		HF	2months	Number of two-month intervals since the creation of the fine
		EF	amClass	Fine amount class: low (amount <50), high (amount $\geq$50)
	Reward	attr.	2months	Number of two-month intervals since the creation of the fine
			payType	Type of payment (null, partial, full or appeal)
		The reward is computed for each MDP's state so that the reward of the complete execution corresponds to the value of the KPI$_{\text{FINES2015}}$ for that execution.		

[a] The component of the reward for an MDP state s related to the interest of the bank is multiplied by a coefficient $c(n) = \frac{(n/\lambda)^2}{1+(n/\lambda)^2}$ that depends on the number of occurrences n of the event log traces that pass through the specific MDP edge with outgoing state s. c goes to 1 when n grows. Here λ is a parameter that can be opportunely tuned, we selected $\lambda = 3$ which corresponds to the median number of edge occurrences in the MDP. This factor is needed to discourage during the RL training the exploitation of some actions that have a positive reward but have low statistic reliability.

Table 3. Example of the transformation of a trace in the corresponding MDP components.

Trace			Enriched trace				MDP		
Activity	Timestamp	Amount	2months	amClass	Actor	payType	Action	Next *state*	Reward
Create fine	13/1/21	40	0	Low	Agent	–	Create fine-0	⟨Create fine, 0, low⟩	0
Send fine	24/1/21	40	0	Low	Agent	–	Send fine-0	⟨Send fine, 0, low⟩	0
Add penalty	18/3/21	60	1	High	Agent	–	Add penalty-1	–	–
Payment	25/7/21	60	3	High	Env.	Full	–	⟨Payment, 3, high⟩	2

Table 4. Results of the simulation evaluation for the *Loan request handling* and the *Traffic fine management* scenarios.

Scenario	Splitting criterion	Policy	Avg. KPI	Offer accepted/full Payment
Loans	60%–40%	RANDOM	36.8	1.4%
		CUSTOMARY	1497.5	38.9%
		OPTIMAL	1727	53.7%
	80%–20%	RANDOM	35.7	1.5%
		CUSTOMARY	1710.5	43.5%
		OPTIMAL	1965.1	61.7%
Fines	60%–40%	RANDOM	1.02	36.9%
		CUSTOMARY	1.12	41.7%
		OPTIMAL	1.17	42.3%
	80%–20%	RANDOM	0.84	30.0%
		CUSTOMARY	0.97	35.0%
		OPTIMAL	1.05	37.1%

KPI of interest for the traces in the test event log that follow the optimal policy (from the first activity) (i) with the value of the KPI of interest of all the traces in the event log, and (ii) with the value of the KPI of interest of the traces in the event log that do not follow the recommended optimal policy. For **RQ2**, we focus on the policy recommended for ongoing executions, i.e., when some activity has already been executed. We hence consider, for each trace in the test event log, all its prefixes and separately analyze each of them, as a potential ongoing execution. For each prefix p of a trace t in the test event log we compare the value of the KPI of interest of the trace t once completed against an estimation of the value of the KPI obtained following the optimal policy from that execution point forward. The estimation is obtained by averaging the KPI values of the traces in the log that have the same prefix as the reference prefix p and follow the optimal policy from there on.

5.3   Results

In this section we report the results of the two scenarios related to the event logs described in Sect. 5.1. For both scenarios, as described in Sect. 5.2, we show (i) the results related to the evaluation of the complete optimal policy (**RQ1**) by reporting first the simulation evaluation and then the test log evaluation; and (ii) the results related to the evaluation of the optimal policy on the test log assuming that some events have already been executed (**RQ2**).

Research Question **RQ1**. Table 4 reports the results related to the simulation evaluation for both the *Loan request handling* and the *Traffic fine management* scenarios. For both splitting criterion (60%–40% and 80%–20%) and for each policy analysed, the average KPI value is displayed together with the percentage of executions for which the bank offer has been accepted by the customer (or the fines have been fully paid by the offender). The policies analysed are: the random policy (RANDOM), the policy selecting the most frequent action in the log for each state (CUSTOMARY) and the optimal (OPTIMAL) policy.

The rows related to the *Loan request handling* scenario (*Loans*) show that for both splitting criteria, the optimal policy (OPTIMAL) generates an average KPI value much higher than the one obtained with a random policy (RANDOM), but also higher than the one obtained with a policy characterized by frequently taken actions (CUSTOMARY). This result confirms that the proposed OPTIMAL policy actually outperforms the policy that is frequently taken in the actual traces, which is considered to be an "optimal" policy by the target agent. Different optimal (and CUSTOMARY) policies are returned with different splitting criteria. When learning with a larger training set and simulating on a smaller test set, the average KPI value increases, for the OPTIMAL and the CUSTOMARY policy, while slightly decreases for the RANDOM policy. Moreover, the table also shows the percentage of traces that, based on the policy simulations, are finally accepted by the customer. By changing the data splitting criteria, the effect is similar to that observed for the average KPI value for the OPTIMAL and the CUSTOMARY policy, with a percentage of accepted offers raising from around 39% to 43% for the CUSTOMARY policy and from around 53% to more than 60% with the OPTIMAL policy. An almost null increase is observed instead for the RANDOM policy.

The results related to the *Traffic fine management* scenario are similar to the results of the loan scenario, as shown in the row *Fines* of Table 4. As for the loan scenario, also in this case, for both splitting criteria, the optimal policy returns higher average KPI values (and hence lower money waste) and produces a higher percentage of traces with fully paid fines than the RANDOM and the CUSTOMARY policies. Also in this case, the difference between the OPTIMAL and the CUSTOMARY policy confirms that the proposed recommendation policy improves the policy actually used in practice. In this scenario, however, the difference in terms of percentage of traces for which fines have fully been paid between the optimal and the random policy is lower than for the *Loan request handling* case. This is possibly due to the overall higher percentage of traces in the FINES2015 event log for which the fines have been fully paid (40%) with respect to the percentage of traces in the BPI2012 log for which the loan offer has been accepted by the customer (17%), as well as to the higher number of actions of $MDP_{FINES2015}$ with respect to the number of actions of $MDP_{BPI2012}$. Moreover, differently from the *Loan request handling* scenario, there is an overall decrease in terms of average KPI value and of traces with fully paid fines when using a larger training set and a smaller test set (80%–20% splitting criterion).

Table 5 shows the results related to the test log evaluation. For each of the two scenarios and for each splitting criterion, we report the number of traces, the average KPI value, as well as the percentage of traces for which the offer has been

Table 5. Results related to the test log evaluation for the *Loan request handling* and the *Traffic fine management* scenario.

Scenario	Splitting criterion	Traces	Trace #	Avg KPI	Offer accepted/full Payment
Loans	60%–40%	ALL	5197	583.3	16.1%
		OPTIMAL P.	1384 (26.6%)	1249.7	34.4%
		NON-OPTIMAL P.	3813 (73.4%)	341.5	9.4%
	80%–20%	ALL	2600	537.2	14.3%
		OPTIMAL P.	753 (29%)	1082.2	30.4%
		NON-OPTIMAL P.	1847 (71%)	315.1	7.7%
Fines	60%–40%	ALL	59946	1.11	40.5%
		OPTIMAL P.	22665 (37.8%)	2.68	90.9%
		NON-OPTIMAL P.	37281 (62.2%)	0.15	9.9%
	80%–20%	ALL	29973	0.96	34.7%
		OPTIMAL P.	9243 (30.8%)	2.76	92.7%
		NON-OPTIMAL P.	20730 (69.2%)	0.16	8.9%

accepted (or the fines have been fully paid) for (i) all traces in the test set (ALL), (ii) the traces in the test set that follow the optimal policy (OPTIMAL P.); (iii) the traces in the test set that do not follow the optimal policy (NON-OPTIMAL P.).

The results of the test log evaluation for the *Loan request handling* scenario (*Loans*) confirm the results obtained with the simulation evaluation. For both splitting criteria, indeed, the average KPI value of the traces following the optimal policy (OPTIMAL P.) is higher than the average KPI value of all the traces (ALL), which in turn is higher than the average KPI value of the traces that do not follow the optimal policy (NON-OPTIMAL P.). The traces following the optimal policy generate an average bank profit of more than 500 euros higher than the average bank profit of all the traces in the event log, as well as of more than 750 euros higher than the average bank profit of the traces that do not follow the optimal policy. The same ranking is obtained if the compared approaches are ordered by the percentage of traces for which the offer by the bank has been accepted by the customer: around 30% for the traces following the optimal policy, around 15% for all traces, and less than 10% for the traces not following the optimal policy. No major differences can be observed between the two splitting criteria, except for a small decrease of the average KPI value and of the percentage of accepted offers.

Similarly to the *Loan request handling* scenario, also in the *Traffic fine management* scenario (rows *Fines* in Table 5) the results of the test log evaluation confirm the findings of the simulation evaluation. Indeed, for both splitting criteria, the traces following the optimal policy (OPTIMAL P.) obtain an average KPI value higher than the average KPI value of all the traces (ALL), which in turn is higher than the average KPI value of the traces that do not follow the optimal policy (NON-OPTIMAL P.). The traces following the optimal policy can produce an average credit value of more than 1 credit higher than the average credit value of all the traces in the event log, as well as of more than 2 credits higher than the average credit value of the traces that do not follow the optimal policy. The trend is also similar for the percentage of traces for which the fine is fully paid. Around 90% of the traces that follow the optimal policy are able to get fully paid fines for both the splitting criteria. While, as in the *Loan request handling*, the percentage of traces with a fully paid fine decreases from the 40% test event log to the 20% event log for the ALL and NON-OPTIMAL P. policies, for OPTIMAL P. the percentage of traces for which the full payment is received is higher for the 20% than for the 40% test event log.

The above results of the two scenarios clearly show that, when no activity has been executed before the target agent starts following the recommendations, the sequence of next activities suggested by the optimal policy generates an average value for the KPI of interest higher than a random policy and than a policy following the most frequently taken actions and, on average, higher than the average KPI value obtained by the actual executions in the test event log (**RQ1**). No clear trends can be observed for different splitting criteria.

Research Question RQ2. As described in Sect. 5.2, we also evaluate the optimal policies at different prefix lengths, that is, by assuming that a part of the execution has already been carried out, before the target agent starts adopting the optimal policy. Figure 2 and Fig. 3 show the average delta KPI value for each prefix length, as well as the prefix occurrence per prefix length. The delta KPI value for each trace and prefix length is computed as the difference between the KPI value obtained by following the optimal policy from that prefix on and the KPI value of the complete trace related to that prefix.

The plot corresponding to the *Loan request handling* scenario (Fig. 2a) shows that for both splitting criteria and for prefix lengths up to 18 there is an average positive delta KPI value, while for longer prefixes a negative or almost null average KPI values are observed. These results can be explained by the low number of traces with length higher than 18 in the test event logs, as it is shown in Fig. 2b.

In the *Traffic fine management* scenario, the plot in Fig. 3a shows a relatively high delta average KPI value for short prefixes (prefixes of length 1 and 2), while the average delta KPI value starts decreasing for traces of prefix length 3. Also in this case, as for the other scenario, the decrease in terms of delta KPI value is mainly due to an overall decrease of the number of traces after prefix 3 (see Fig. 3b). Differently from the *Loan request handling* scenario, as already observed during the discussion of **RQ1**, the average delta KPI value obtained with the 80%–20% splitting criterion is higher than the one obtained with the 60%–40% splitting criterion, except that for prefix length 3.

In conclusion, these results confirm that even when considering ongoing executions, the recommended sequence of next activities suggested by the proposed optimal policy generates higher average KPI values than the ones obtained by actual executions in the test event log (**RQ2**).

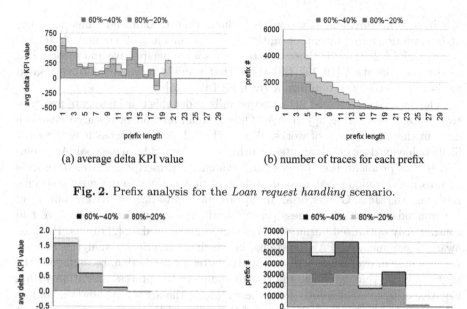

(a) average delta KPI value   (b) number of traces for each prefix

Fig. 2. Prefix analysis for the *Loan request handling* scenario.

(a) average delta KPI value   (b) number of traces for each prefix

Fig. 3. Prefix analysis for the *Traffic fine management* scenario.

Beyond the performance perspective, we briefly comment here on the plausibility of the optimal policies obtained. The major contributions of the policies for the two cases are clear and reasonable. In the *Loan request handling* scenario the policy advises to accept more loan applications, so as to increase the number of possible accepted loans. Moreover, it advises to increase the interaction between the bank and the customer, with the creation of multiple offers and the subsequent call to the customer. In the *Traffic fine management* scenario the policy advises to send the fine early to the offender, so as to raise the probability that he/she pays the fine on time.

6   Related Work

The state-of-the-art works related to this paper pertain to two fields: Prescriptive Process Monitoring and Reinforcement Learning. The section is hence structured by first presenting Prescriptive Process Monitoring related works and then Reinforcement Learning state-of-the-art works, applied to process mining problems.

Several Prescriptive Process Monitoring techniques have been recently proposed in the literature. Focusing on the type of interventions that the approaches recommend [12], we can roughly classify existing work in Prescriptive Process

Monitoring in three main groups: (i) those that recommend different types of interventions to prevent or mitigate the occurrence of an undesired outcome [8,17,18,20,24]; (ii) those that take a resource perspective and recommend a resource allocation [19,22]; (iii) those that provide recommendations related to the next activity to optimize a given KPI [2,9,25].

The approach presented in this paper falls under this third family of prescriptive process monitoring approaches. Only a small amount of research has been done in this third group of works. Weinzierl et al. in [25] discuss how the most likely behavior does not guarantee to achieve the desired business goal. As a solution to this problem, they propose and evaluate a prescriptive business process monitoring technique that recommends next best actions to optimize a specific KPI, i.e., the time. Gröger et al. in [9] present a data-mining driven concept of recommendation-based business process optimization supporting adaptive and continuously optimized business processes. De Leoni et al. in [2] discuss Process-aware Recommender (PAR) systems, in which a prescriptive-analytics component, in case of executions with a negative outcome prediction, recommends the next activities that minimize the risk to complete the process execution with a negative outcome. Differently from these state-of-the-art works, however, in this work we take the perspective of one of the actors of the process and we aim at optimizing a domain-specific KPI of interest for this actor by leveraging an RL approach.

In the literature, only few RL approaches have been proposed for facing problems in the process mining field. Silvander proposes using Q-Learning with function approximation via a deep neural network (DQN) for the optimization of business processes [21]. He suggests defining a so called decay rate to reduce the amount of exploration over time. Huang et al. employ RL for the dynamic optimization of resource allocation in business process executions [11]. Metzger et al. propose an alarm-based approach to prevent and mitigate an undesired outcome [17]. They use online RL to learn when to trigger proactive process adaptations based on the reliability of predictions. Although all these works use RL in the process mining field, none of them use it for recommending the next actions to perform in order to optimize a certain KPI of interest, as in this work.

Finally, some works have applied RL and Inverse Reinforcement Learning (IRL) approaches to recommend the next actions on temporal data [16] or on data constrained by temporal constraints [1].

7   Conclusion

In this paper we have proposed the use of RL in the solution of the problem of computing next activity recommendations in Prescriptive Process Monitoring problems.

Differently from other state-of-the-art works our model handles non deterministic processes, in which only part of the activities are actually actionable and the rest of them are, from the target actor point of view, stochastically selected by the system environment. This is a common situation in multi-actors processes. By taking the decision making perspective of one of the actors involved in a process (target actor), we first learn from past executions the behaviour of the environment and we then use RL to recommend the best activities to carry on in order to optimize a measure of interest. The obtained results show the goodness of the proposed approach in comparison to the policy used by the actor, i.e., without using recommendations.

We plan to extend this approach by including in the MDP state the raw information related to the history of the process execution, so as to automate as much as possible the pre-processing phase of our computational pipeline. However, in that case the consequent increase of the state space dimension and its cardinality would require the usage of state generalisation techniques, such as, those implemented with Deep Reinforcement Learning or by applying smart clustering techniques. Moreover, we would like to explore the possibility to use declarative constraints for defining and enforcing domain knowledge constraints.

References

1. De Giacomo, G., Iocchi, L., Favorito, M., Patrizi, F.: Foundations for restraining bolts: Reinforcement learning with LTLF/LDLF restraining specifications. In: Proceedings of the 29th International Conference on Automated Planning and Scheduling, ICAPS 2018, pp. 128–136. AAAI Press (2019)
2. de Leoni, M., Dees, M., Reulink, L.: Design and evaluation of a process-aware recommender system based on prescriptive analytics. In: 2nd International Conference on Process Mining (ICPM 2020), pp. 9–16. IEEE (2020)
3. Di Francescomarino, C., Ghidini, C., Maggi, F.M., Milani, F.: Predictive process monitoring methods: which one suits me best? In: Weske, M., Montali, M., Weber, I., vom Brocke, J. (eds.) BPM 2018. LNCS, vol. 11080, pp. 462–479. Springer, Cham (2018). https://doi.org/10.1007/978-3-319-98648-7_27
4. van Dongen, B.: Bpi challenge 2012, April 2012. https://doi.org/10.4121/uuid:3926db30-f712-4394-aebc-75976070e91f
5. van Dongen, B.: Bpi challenge 2017, February 2017. https://doi.org/10.4121/uuid:5f3067df-f10b-45da-b98b-86ae4c7a310b
6. Dumas, M.: Constructing digital twins for accurate and reliable what-if business process analysis. In: Proceedings of the International Workshop on BPM Problems to Solve Before We Die (PROBLEMS 2021). CEUR Workshop Proceedings, vol. 2938, pp. 23–27. CEUR-WS.org (2021)
7. Dumas, M., et al.: Augmented business process management systems: a research manifesto. CoRR abs/2201.12855 (2022). https://arxiv.org/abs/2201.12855
8. Fahrenkrog-Petersen, S.A., et al.: Fire now, fire later: alarm-based systems for prescriptive process monitoring. arXiv preprint arXiv:1905.09568 (2019)
9. Gröger, C., Schwarz, H., Mitschang, B.: Prescriptive analytics for recommendation-based business process optimization. In: International Conference on Business Information Systems. pp. 25–37. Springer (2014). https://doi.org/10.1007/978-3-319-06695-0_3
10. Hu, J., Niu, H., Carrasco, J., Lennox, B., Arvin, F.: Voronoi-based multi-robot autonomous exploration in unknown environments via deep reinforcement learning. IEEE Trans. Vehicul. Technol. **69**(12), 14413–14423 (2020)
11. Huang, Z., van der Aalst, W., Lu, X., Duan, H.: Reinforcement learning based resource allocation in business process management. Data Know. Eng. **70**(1), 127–145 (2011)

12. Kubrak, K., Milani, F., Nolte, A., Dumas, M.: Prescriptive process monitoring: Quo vadis? CoRR abs/2112.01769 (2021). https://arxiv.org/abs/2112.01769
13. de Leoni, M.M., Mannhardt, F.: Road traffic fine management process (2015). https://doi.org/10.4121/uuid:270fd440-1057-4fb9-89a9-b699b47990f5
14. Maggi, F.M., Di Francescomarino, C., Dumas, M., Ghidini, C.: Predictive monitoring of business processes. In: Jarke, M., et al. (eds.) CAiSE 2014. LNCS, vol. 8484, pp. 457–472. Springer, Cham (2014). https://doi.org/10.1007/978-3-319-07881-6_31
15. Mannhardt, F., de Leoni, M., Reijers, H.A., van der Aalst, W.M.P.: Balanced multi-perspective checking of process conformance. Computing **98**(4), 407–437 (2015). https://doi.org/10.1007/s00607-015-0441-1
16. Massimo, D., Ricci, F.: Harnessing a generalised user behaviour model for next-poi recommendation. In: Proceedings of the 12th ACM Conference on Recommender Systems, RecSys 2018, pp. 402–406. ACM (2018)
17. Metzger, A., Kley, T., Palm, A.: Triggering proactive business process adaptations via online reinforcement learning. In: International Conference on Business Process Management, pp. 273–290. Springer (2020). https://doi.org/10.1007/978-3-030-58666-9_16
18. Metzger, A., Neubauer, A., Bohn, P., Pohl, K.: Proactive process adaptation using deep learning ensembles. In: International Conference on Advanced Information Systems Engineering, pp. 547–562. Springer (2019). https://doi.org/10.1007/978-3-030-21290-2_34
19. Park, G., Song, M.: Prediction-based resource allocation using LSTM and minimum cost and maximum flow algorithm. In: International Conference on Process Mining, ICPM 2019, pp. 121–128. IEEE (2019)
20. Shoush, M., Dumas, M.: Prescriptive process monitoring under resource constraints: a causal inference approach. CoRR abs/2109.02894 (2021). arxiv.org/abs/2109.02894
21. Silvander, J.: Business process optimization with reinforcement learning. In: Shishkov, B. (ed.) BMSD 2019. LNBIP, vol. 356, pp. 203–212. Springer, Cham (2019). https://doi.org/10.1007/978-3-030-24854-3_13
22. Sindhgatta, R., Ghose, A., Dam, H.K.: Context-aware analysis of past process executions to aid resource allocation decisions. In: Nurcan, S., Soffer, P., Bajec, M., Eder, J. (eds.) CAiSE 2016. LNCS, vol. 9694, pp. 575–589. Springer, Cham (2016). https://doi.org/10.1007/978-3-319-39696-5_35
23. Sutton, R.S., Barto, A.G.: Reinforcement Learning: An Introduction. The MIT Press, 2nd edn. (2018)
24. Teinemaa, I., Tax, N., de Leoni, M., Dumas, M., Maggi, F.M.: Alarm-based prescriptive process monitoring. In: Weske, M., Montali, M., Weber, I., vom Brocke, J. (eds.) BPM 2018. LNBIP, vol. 329, pp. 91–107. Springer, Cham (2018). https://doi.org/10.1007/978-3-319-98651-7_6
25. Weinzierl, S., Dunzer, S., Zilker, S., Matzner, M.: Prescriptive business process monitoring for recommending next best actions. In: Fahland, D., Ghidini, C., Becker, J., Dumas, M. (eds.) BPM 2020. LNBIP, vol. 392, pp. 193–209. Springer, Cham (2020). https://doi.org/10.1007/978-3-030-58638-6_12

Predictive Process Monitoring

Change Detection in Dynamic Event Attributes

Jonas Cremerius(✉) and Mathias Weske

Hasso Plattner Institute, University of Potsdam, Potsdam, Germany
{jonas.cremerius,mathias.weske}@hpi.de

Abstract. Discovering and analysing business processes are important tasks for organizations. Process mining bridges the gap between process management and data science by discovering process models using event logs derived from real-world data. Besides mandatory event attributes like case identifier, activity, and timestamp, additional event attributes can be present, such as human resources, costs, and laboratory values. These event attributes can be modified by multiple events in a trace, which can be classified as so-called *dynamic* event attributes. So far, the process behaviour of event attributes is described in the form of read/write operations or object-lifecycle states. However, the actual value behaviour has not been considered yet. This paper introduces an approach that allows to automatically detect changes in the actual values of *dynamic* event attributes, enabling to identify changes between process activities representing events with the same activity name. This can help to confirm expected behaviour of *dynamic* event attributes, but also allows deriving novel insights by identifying unexpected changes. We applied the proposed technique on the MIMIC-IV real-world data set on hospitalizations in the US and evaluated the results together with a medical expert. The approach is implemented in Python with the help of the PM4Py framework.

Keywords: Process mining · Change detection · Process enhancement

1 Introduction

Businesses organizations seek to find valuable insights out of data stored in information systems with the aim to improve their business processes. Today, such information systems can include data about end-to-end processes or even beyond that. Due to that fact, process mining was developed to understand the actual execution of business processes, providing techniques for process discovery, conformance checking, and enhancement [1].

In process discovery, the discovered process model can be analysed based on the occurred events, their order, and frequency. Event logs might contain additional data, so-called event attributes, providing further information about an event, which can be used to enhance process models [10].

© Springer Nature Switzerland AG 2022
C. Di Ciccio et al. (Eds.): BPM 2022, LNBIP 458, pp. 157–172, 2022.
https://doi.org/10.1007/978-3-031-16171-1_10

Event attributes can be *dynamic* in the sense that they are stored in multiple events, such as an order status or laboratory values, which evolve through the process. As *dynamic* event attributes occur multiple times during the process, understanding their development can be of interest [11]. So far, the process behaviour of event attributes is described in the form of read/write operations or object-lifecycle states [6, 16].

However, there is still a lack of describing the actual value behaviour of *dynamic* event attributes. For example, it might be of interest to see if a stay in an intensive care unit (ICU) results in improved laboratory values of a patient in the recovery ward. Thus, we can compare the laboratory values conducted in the ICU to the ones in the recovery ward.

Therefore, this paper provides an approach to automatically detect changes in *dynamic* event attributes, so that it is not only known if the values change throughout the process, but also at which activity representing all events with the same activity name and in which direction (increasing, decreasing).

The remainder of this paper is organized as follows. Section 2 provides related work, and Sect. 3 introduces preliminaries. Section 4 presents the approach for change detection in *dynamic* event attributes, and Sect. 5 applies the approach to the MIMIC-IV real-world data set on hospitalizations. We discuss the approach and its limitations in Sect. 6 before the paper is concluded in Sect. 7.

2   Related Work

The analysis of event attributes has been approached from different perspectives in the literature.

A prominent application is the identification of decision rules, such as in data-aware heuristic mining [16]. Regarding the exploration of event attributes, the multi-perspective process explorer allows investigating the distribution of each event attribute at each activity [18]. Data-enhanced process models add aggregated information about event attributes, such as the mean value, to the process model activities representing the events. In data-enhanced process models, the selection of event attributes for detailed analysis is supported by allowing filtering according to their process behaviour and the degree of variability through the process [11]. In [6], the access to event attributes is described and annotated to the process model, describing the data object lifecycle of each event attribute.

While there exist approaches trying to better explore and understand the actual values of event attributes, there remains, to our knowledge, a lack of understanding the changing behaviour of these values. The work describing the data object lifecycle is already a step in this direction, but lacks support for understanding the change of the actual values behind the event attributes.

Change detection is highly present in time series data, which refers to the problem of finding abrupt changes in data when a property of the time series changes [4]. In terms of process analysis, change detection has been applied to detect and explain concept drifts. In [2], event attributes are used to explain concept drifts, such as that a decrease in the age of customers led to an increase in the prevalence of the email notification activity.

However, time series change detection accepts only one value per time point, which requires methods of aggregations when analysing groups, which is the typical use case in process mining. This leads to information loss and lacks a detailed representation of the analysed group [4].

To overcome this limitation, statistical tests allow comparing two timestamps in more detail. For example, the Wilcoxon Signed-rank Test considers all values of the analysed group and ranks the differences between two timestamps to answer the question, if there is a statistically significant change [15]. This form of change detection is popular in the medical domain, where before-after comparisons are conducted. For example, [9] compares a laboratory value measured at inpatient admission and 72 h after that.

In process mining, statistical tests are used to retrieve a variety of insights. For example, the difference of event durations is assessed between two groups in an emergency process [12], which is not a before-after comparison, but still compares the difference of values in two groups. The same holds for process variant comparison, where the event transition frequency is compared between two process variants [19].

In this contribution, we propose to use statistical tests to detect changes of event attributes in the process. In particular, we make use of the before-after comparison of statistical tests to detect changes of dynamic event attributes between process activities, which has not been conducted in process mining so far to our knowledge.

3 Preliminaries

This paper builds on the contribution of Supporting Domain Data Selection in Data-Enhanced Process Models [11], which starts with an event log. An event log consists of sequences of events, which are grouped into traces. An event can have an arbitrary number of additional event attributes. The following definition is based on [17].

Definition 1 (Event log, Trace, and Event). Let V be the universe of all possible values and E_A be the universe of event attributes. An event e is a mapping of event attributes to values, such as $e \in E_A \to V$. The universe of events is defined as $E_U = E_A \to V$. If an event $e \in E_U$ has no value assigned to an event attribute $e_{At} \in E_A$, it is denoted as $e(e_{At}) = \perp$. A trace $t \in E_U{}^*$ is a sequence of events, and $T \subseteq E_U{}^*$ represents the respective universe of traces, in which all events are unique. An event log L is a set of traces, so $L \subseteq T$, where each trace is unique as well. As events and traces are unique, we say, that two traces $t_1, t_2 \in L$ belong to the same trace variant $t_{Var} \subseteq L$, if the events in the traces have the same activity ordering and number of events. We refer to T_{Var} as the universe of trace variants.

Normally, an event represents an activity which is conducted within a certain case at a given time, represented by a timestamp. These are treated as regular event attributes in this contribution, so we assume activity, case, and timestamp.

The event instances of a given trace are ordered by their timestamp and have the same case. For simplicity, we assume that the timestamps of events in a trace are never equal. We further assume, that the data type of one event attribute is always the same for all events.

Given events $e_i \neq e_j$ in a given trace t_i, let $e_i > e_j$ represent a directly follows relationship, if e_j appears after e_i and there does not exist an event e_k in t_i which appears between e_i and e_j. Let $e_i \twoheadrightarrow e_j$ represent an eventually follows relationship, if e_j appears at any position after e_i in t_i.

Event attributes can be classified according to their process behaviour, which is based on [11].

Definition 2 (Event Attribute Classification). Before an event attribute $e_{At} \in E_A$ can be classified, the activities using the event attribute and the average number of events using it per trace need to be identified.

Given an event attribute $e_{At} \in E_A$ in an event log L, the set $e_{At_{Act}}$ represents all activities in which the event attribute is used.

$$e_{At_{Act}} := \{e(activity) \in V \mid e(e_{At}) \neq \bot, e \in t, t \in L\} \tag{1}$$

With that, it is known which activities have an event attribute, but it remains unclear whether an event attribute is changing during the process. Therefore, $e_{At_{AvgTrace}}$ describes the average number of events having the event attribute per trace. First, the event log is filtered, so that only traces are included which use the event attribute at least once:

$$L_{e_{At}} = \{t \in L \mid (\exists e \in t)[e(e_{At}) \neq \bot]\} \tag{2}$$

Then, the average number of occurrences of the event attribute per trace can be calculated:

$$e_{At_{AvgTrace}} = \frac{\sum_{t \in L_{e_{At}}} \sum_{e \in t}[e(e_{At}) \neq \bot]}{|L_{e_{At}}|} \tag{3}$$

Three different process characteristics (pc) are defined based on the previously defined features $|e_{At_{Act}}|$ and $e_{At_{AvgTrace}}$.

$$pc(e_{At}) = \begin{cases} static, & |e_{At_{Act}}| = 1, e_{At_{AvgTrace}} = 1 \\ semi-dynamic, & |e_{At_{Act}}| > 1, e_{At_{AvgTrace}} = 1 \\ dynamic, & |e_{At_{Act}}| \geq 1, e_{At_{AvgTrace}} > 1 \end{cases}$$

4   Approach

In this contribution, the goal is to describe the changing behaviour of *dynamic* event attributes through a process. To clarify that, Table 1 illustrates an example event log mimicking a hospital process. Besides the mandatory entries, it contains laboratory values in the form of event attributes. As these are associated to multiple activities and occur multiple times per trace, these are classified as *dynamic* event attributes. Thus, these are suited for the analysis steps proposed in this paper.

Before the approach is presented, we clarify what kind of change we intend to detect. Our idea is to bring meaning behind the timestamps in the form of activity names and allow identifying, how activities potentially influence the values of event attributes. Therefore, we say that an event attribute changes not if it changes at an arbitrary point of time, but when there is a change in the values between activities. On top of this, we want to achieve this by considering all values of the respective activities.

Table 1. Example event log describing a high level hospital process having laboratory values as event attributes

Case ID	Activity	Timestamp	Bicarbonate value	Creatinine value
1	Admit to hospital	1	140	0.7
1	Treat in medical ward	2	200	0.7
1	Discharge patient	3	120	0.8
2	Admit to hospital	1	135	0.6
2	Treat in ICU	2	100	0.6
2	Discharge patient	3	150	0.7

Looking at the example event log in Table 1, there is a difference in the development of the Bicarbonate laboratory value, dependent on which ward is visited during the hospital process. While it increases in the "Treat in Medical Ward" activity, it decreases in the "Treat in ICU" ward. In the following approach, we identify these changes not in single traces, but make statements for all traces in the event log, deriving a common behaviour of *dynamic* event attributes in the process.

4.1 The Three Dimensions of Change

In this contribution, a three-dimensional perspective is suggested to identify changes in *dynamic* event attributes, which is illustrated in Fig. 1.

The first dimension on the x-axis is the event attribute, because it is the goal of this paper to understand the behaviour of event attributes. The second dimension on the y-axis shows all directly follows and eventually follows relations in the event log, which represent the points of change in the process. Lastly, the z-axis adds information about changes in trace variants, which provides additional context to the relation information on the y-axis. This information is important to preserve the process context, as it might be the case that the process before and after any relation might have an influence on the behaviour of an event attribute.

We start formalizing this construct by defining a change detection cube:

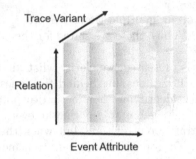

Fig. 1. The three dimensions of change

Definition 3 (Change Detection Cube). We define a change detection cube (CDC_L) for a given event log $L \subseteq T$ as a set of change analysis cells (cac), such that $CDC_L := E_{A_L} \times DFR_L^+ \times T_{Var_L}$, where $E_{A_L} \subseteq E_A$ is the set of event attributes being assigned a value $\neq \perp$ at any event $e \in t, t \in L$ and $DFR_L^+ \subseteq V \times V$ is the transitive closure of directly follows relationships, such that it contains the eventually follows relationships as well. The elements of DFR_L^+ consist of the respective activity names, so if $e_i > e_j$, $(e_i(activity), e_j(activity)) \in DFR_L^+$ and if $e_i \rightarrow e_j$, $(e_i(activity), e_j(activity)) \in DFR_L^+$. $T_{Var_L} \subseteq T_{Var}$ refers to the set of all trace variants in L.

A change analysis cell $cac \in CDC_L$ represents one cell in the cube, such that $cac = (e_{At} \in E_{A_L}, rel \in DFR_L^+, t_{Var} \in T_{Var_L})$. One cell in the cube refers to a single change detection, for example, looking at Table 1, the Bicarbonate value between the activities "Treat in Medical Ward" and "Discharge Patient" decreases in a trace variant in which the activity "Admit to hospital" is included.

The idea of analysing three-dimensional data in a cube perspective goes back to on-line analytical processing (OLAP), where so-called OLAP cubes were introduced, which can be of higher dimensions as well [8]. These allow operations, which can be applied on the change detection cube:

- Slice: Reduces the cube to a two-dimensional view by selecting a specific value for one dimension, such as the analysis of all changes for one event attribute
- Dice: Creates a sub-cube where specific values for all dimensions can be specified, e.g., analyse all changes for a subset of event attributes
- Pivot: Rearranges the dimensions, such that event attributes and relations swap their axis
- Drill up/down: Changes the level of aggregation in the dimensions, e.g., trace variants could be merged together

With CDC_L defined, each element $cac \in CDC_L$ refers to a change detection analysis, which is defined next.

Definition 4 (Change Detection Analysis). Given an event log $L \subseteq T$ with its respective change detection cube CDC_L, we define a change detection analysis

(CDA) as a function mapping each $cac \in CDC_L$ to a pair of two values representing the result of the change analysis, such that $CDA_L = CDC_L \rightarrow V \times V$.

The result of the change analysis consisting of a two-value pair is generated by statistical tests, which are described next.

4.2   Change Detection as a Before-After Comparison

Given a change detection cube CDC_L for a given event log $L \subseteq T$, we propose to detect changes for each change analysis cell $cac \in CDC_L$ with its elements $e_{At} \in E_{A_L}$, $rel \in DFR_L^+$, and $t_{Var} \in T_{Var_L}$. The relation rel in a cell consists of two activity names (a_1, a_2) of which the events are in a directly follows or eventually follows relationship.

To detect changes, we need to derive the respective event attribute values of e_{At} for both activities (a_1, a_2) from the trace variant $t_{Var} \in T_{Var_L}$. For that, we define a multiset EAV_{cac} for each change analysis cell cac, in which the elements consist of event attribute value pairs $(e_i(e_{At}), e_j(e_{At}))$ with $e_i(activity) = a_1$, $e_j(activity) = a_2$, and $e_i(e_{At}), e_j(e_{At}) \neq \bot$, where the traces including the events e_i, e_j are in the respective trace variant $t_{Var} \subseteq L$. If the respective events are directly following, we only consider directly follows relations in the traces, as it could be the case that a trace includes the directly follows relationship and at some point an eventually follows relationship of both events. Thus, there can be cases where a separate analysis of directly and eventually follows relations makes sense, which could be solved by treating these as separate relations in DFR_L^+, where one is the directly follows and the other the eventually follows relation.

Further, this approach might lead to multiple entries for one case, if the trace includes loops containing the same directly follows or eventually follows relationship. This is intended, as we are interested in the changing behaviour between both activities. However, it could be interesting to investigate the looping behaviour in more detail, such that a value tends to change in the first occurrence of the relation, but remains constant after that. This could be implemented by adding a loop index to each change analysis cell, resulting in separate change analysis cells for each loop iteration. For example, if the relation (a, b) occurs twice in a trace, one could analyse the changing behaviour for the first and second occurrence of (a, b) separately.

With EAV_{cac} representing event attribute value changes for a change analysis cell $cac \in CDC_L$, there exist multiple event attribute values for both activities, given that there are multiple traces related to the change analysis cell. Understanding the changing behaviour between two sets of values is a typical use case in the field of statistical analysis, especially before-after comparisons, e.g., the comparison of laboratory values between two timestamps [9]. As this approach investigates the behaviour of directly follows and eventually follows relationships, we can perform such a before-after comparison for each change analysis cell $cac \in CDC_L$.

We will now introduce statistical tests used for comparing event attribute values in $cac \in CDC_L$.

4.3   Statistical Tests

To conduct statistical tests, two hypotheses need to be provided. First, the null hypothesis states that there is no difference between two samples. These two samples are the event attribute values of two activities represented by EAV_{cac}. Thus, the null hypothesis says, that there is no change in the event attribute values. The task of the statistical test is to either reject or confirm the null hypothesis. By rejecting the null hypothesis, the alternative hypothesis, saying that there is a change in the event attribute values, can be confirmed. We can never say that there is a change for each sample taken, but provide a probability that a given result would occur under the null hypothesis [21]. This probability is the p-value. Thus, the lower the p-value, the lower the chance, that a given sample is not changing. That is the reason why a significance threshold α is used to reject the null hypothesis, which is typically 0.05.

If multiple tests are conducted on the same samples, which is the case when multiple event attributes are analysed for the same relation and trace variant, α can be adjusted by performing a Bonferroni correction [5]. For example, if 10 event attributes are under analysis, one would divide α by 10, resulting in $\alpha = 0.005$. We will not determine a concrete α, but suggest using 0.05 with the option to apply Bonferroni correction, as the application of the correction method depends on the analysis goal. For example, if one wants to determine, if there is no change in any event attribute (universal null hypothesis), the correction should be applied [5].

Choosing the appropriate statistical test is based on three factors. The first factor is the event attribute type, which is either continuous or categorical. We will use the method proposed in [11] to identify the variable type of event attributes in event logs by comparing the total number of values vs. the amount of unique values of a variable. Second, the distribution of data is important. As we cannot make any assumptions about the distribution of each event attribute, we make use of so-called non-parametric tests. Lastly, the relation between the samples under comparison needs to be considered, which is either paired or unpaired. In our case, we have paired samples, because the event attribute values from both activities come from the same case and are not independent. Considering these factors, we end up with the Wilcoxon Signed-rank Test for continuous event attributes and the Stuart-Maxwell Test for categorical event attributes [20].

Wilcoxon Signed-Rank Test. Given a change analysis cell $cac \in CDC_L$ with its event attribute values EAV_{cac}, the Wilcoxon Signed-rank test performs pairwise comparison of each element $(e_i(e_{At}), e_j(e_{At})) \in EAV_{cac}$, given that e_{At} is continuous. The test makes use of the *Simple Difference Formula*, which results in the difference between the proportion of favourable and unfavourable pairs $RBC = f - u$, the so-called matched-pairs rank-biserial correlation, whereas favourable/unfavourable represent the pairs where the differences have the same sign (increasing or decreasing) [15]. As we do not test for a specific direction, we will speak of increasing/decreasing instead of favourable/unfavourable. Table 2 demonstrates an example, where all pairs in EAV_{cac} are compared according to their difference in the activities specified in cac. The test calculates each difference, which is shown in the "Change" column. Dependent on the degree of change, ranks are assigned, where increasing/decreasing changes are differentiated in the respective column.

Table 2. Wilcoxon Signed-Rank Test example

Case ID	Treat in ICU	Discharge patient	Change	Increasing	Decreasing
1	150	200	50	5	–
2	140	160	20	3	–
3	100	110	10	1	–
4	150	135	−15	–	2
5	150	180	30	4	–
6	200	185	−15	–	2

As mentioned before, the test considers the rank sums, which are 13 for the increasing and 4 for the decreasing pairs. RBC is then the relative difference of both, which is $13/17 - 4/17 = 0.523$. It can take values between -1 and 1, dependent on whether the majority of changes are increasing or decreasing. Thus, it does not only consider if there is a difference in one direction, but also provides information about how many of the major changes go into the respective direction. In combination with a p-value, we can say, that the difference is statistically significant as well.

The major advantage of this test is its simplicity, with its comprehensible calculation of the difference between two groups. Additionally, its result is directional, which automatically identifies an increasing or decreasing behaviour [15].

Stuart-Maxwell Test. If the event attribute e_{At} is categorical, the Stuart-Maxwell test, which is also called the Generalized McNemar test, can be used to identify changing behaviour. In comparison to McNemar, this test can deal with an arbitrary amount of categories [23]. Tests for categorical variables use so-called contingency tables, which represent the transition frequency from one category to the others for before-after comparison. Table 3 illustrates an example of a contingency table of a variable with three categories. It can be seen, for example, that there are 100 cases, where the event attribute remains high and that the event attribute changes from high to normal in 50 cases.

The test checks for so-called *marginal homogeneity*. *Marginal homogeneity* refers to equality between one or more of the row marginal proportions and the corresponding column proportions [23]. For example, the category high in Table 3 has no marginal homogeneity, because the proportion of the row is different to the proportion of the column including the respective category (first row(50) vs. first column(0) without high/high). The test checks this for all categories

Table 3. Contingency table example

–	High	Normal	Low
High	100	50	0
Normal	0	50	25
Low	0	0	75

and results in a p-value p and a chi-squared value χ^2, indicating a change in the respective variable or not, whereas p provides information about statistical significance and χ^2 gives information about how marginal proportions are not homogeneous. Thus, the higher the proportion are not homogeneous, the higher the change in the categories. The exact calculation will not be covered in this paper, but is conducted as described in [23].

The results of the statistical tests of each change analysis cell $cac \in CDC_L$ will be represented as a change detection analyses, such that $CDA_L(cac) = (p, t)$, where the test-statistic t is RBC for continuous event attributes and χ^2 for categorical event attributes.

4.4   Connecting Continuous and Categorical Event Attributes

The differentiation between continuous and categorical event attributes enforces a separate analysis of both. Nevertheless, some event attributes might be connected to each other. A categorical event attribute could describe different states for a continuous event attribute, such as being high, normal, or low. A prominent example are laboratory values, which have these states in addition to their plain value. Thus, there is one attribute for the continuous laboratory value and another one for the categorical laboratory value in the event log. Another example are sensor data, such as temperature measurements etc. Thus, we propose to connect continuous and categorical event attributes by creating a link between change analysis cells $cac \in CDC_L$. This allows to identify, whether a changing behaviour in a continuous event attribute is also represented in the respective categorical event attribute and the other way around. Thus, we define $EAC_L = CDC_L \rightarrow CDC_L$ as an event attribute connection, linking the respective change analysis cells. If there exists no connection, we denote that as $EAC_L(cac) = \bot$.

The linking has to be performed manually, as we do not know of any standardized naming of event attributes in event logs. For example, one could name them equally and assign a variable type to them, which would make the connection trivial.

Next, the proposed approach is evaluated on a real-world healthcare data set, derived from the MIMIC-IV database.

5   Evaluation

The proposed approach was implemented in Python with the help of the PM4Py framework[1] [7]. The relevance of this approach is illustrated in a medical environment, where we generated an event log from the Medical Information Mart for Intensive Care IV (MIMIC-IV) database. The reason for choosing this database is its richness of data, allowing to generate event logs with multiple *dynamic* event attributes.

5.1   Dataset

MIMIC-IV is a relational database including hospital processes of different patients, with procedures performed, medications given, laboratory values taken, image analysis conducted, and more. Its purpose is to support research in healthcare and is therefore publicly available [14].

The event log extracted from MIMIC-IV incorporates a high-level process, describing department visits of patients during their hospital stay, such as emergency department or intensive care unit (ICU). The event log contains 3447 hospital process instances with 13795 events of acute kidney failure (AKF) patients. AKF was chosen together with a medical expert, because of its high prevalence and its measurable disease progression by kidney specific laboratory values.

For each department visit, the event log provides up to 62 event attributes, including laboratory values and demographic information. 56 event attributes represent laboratory values, which are classified as *dynamic*. 28 *dynamic* event attributes are continuous and 28 are categorical. The categorical laboratory values store information about abnormality of the respective continuous value. Thus, we present an event log with multiple *dynamic* event attributes being on different scales with a balance between categorical and continuous event attributes.

5.2   Results

We applied the proposed approach on the event log introduced above. The resulting change detection cube CDC_L can be explored with our artefact. The artefact supports the proposed OLAP operations (Slice, Dice, Pivot, and Drill up/down), where we decided to always slice the cube to enable the exploration of the change analysis results. Therefore, we end up with a two-dimensional event attribute change matrix. Figure 2 illustrates an arbitrary view of CDC_L, showing a subcube with continuous event attributes chosen together with the medical expert and relations having the most changes. The cube was sliced and drilled down to represent all trace variants. Further views, which also consider trace variants, are provided in the already mentioned GitHub repository.

Each cell in the matrix represents one change analysis cell $cac \in CDC_L$ and the number inside displays the test-statistic of the change detection analysis

---

[1] https://github.com/jcremerius/Change-Detection-in-Dynamic-Event-Attributes.

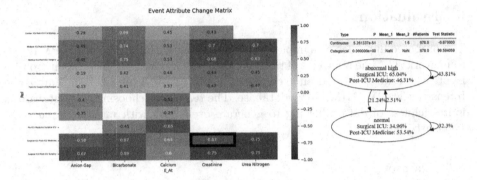

Fig. 2. Change Detection Analysis, illustrating an Event Attribute Change Matrix with the significant event attribute changes and a detailed view of one cell with the connection to the respective categorical event attribute.

$CDA_L(cac)$, which is the RBC value for continuous event attributes. The colour of the cell illustrates the change direction, where blue is decreasing and red is increasing. The cell is blank, if there is no statistical evidence for a change according to the given significance threshold α, which is 0.05 in this case.

The transition between the department visits is shown as relations on the y-axis of the matrix. Emergency department is not listed, as it does not contain any *dynamic* event attributes. The matrix shows, that the laboratory values change differently dependent on the patient's progress through the hospital. For example, we observe no value changes of creatinine between pre-ICU and ICU treatment, whereas it decreases after the ICU stay significantly. On the other hand, the values of calcium tend to decrease in the ICU and increase after that. The developed artefact allows displaying significant changes in a process model, which is shown in Fig. 3, presenting significant changes of calcium.

One can also analyse a change analysis cell in more detail by clicking on it in the matrix, which is shown in Fig. 2 on the right-hand side, where the cell marked with the black box is selected. The figure shows the test results in more detail and illustrates the event attribute connection EAC_L, where the cell of the continuous event attribute "Creatinine" is connected to the respective categorical event attribute "Abnormal Creatinine". The graph nodes show the respective categories, which are "abnormal high" and "normal". The arrows are annotated with the amount of samples changing their state. As the degree of change is high with 21% from "abnormal high" to "normal", the categorical test ended up with a p-value so close to 0 that it is displayed as being 0. This shows the importance of the test-statistic, because the p-value only says, that a change is present, but not how high the degree of change is. Thus, the change in the continuous event attribute results in a change in the categorical event attribute as well.

To verify the attribute value changes, we looked into medical literature and asked a medical expert for consultation. Urea nitrogen and creatinine are estab-

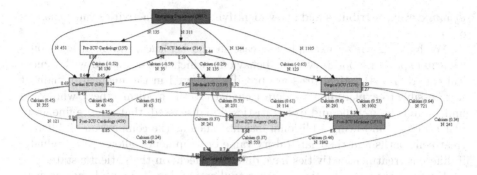

Fig. 3. Directly-Follows Graph enhanced with event attribute changes. The edge labels show the event attribute name with its RBC value and sample size. The colours illustrate the value direction, where blue is decreasing and red is increasing. The ends of the edges show the mean value of the event attribute at the respective activity. (Color figure online)

lished parameters for renal recovery and are expected to decrease after ICU treatment [22]. Additionally, bicarbonate use in the ICU for treatment of anion-gap metabolic acidosis avoids the need for dialysis, which is generally the first-line therapy for acidosis [13]. That explains the increase of bicarbonate and the decrease of anion gap after ICU treatment. The value behaviour of calcium is an interesting observation, as it decreases in the ICU and increases after that, resulting in no significant change between pre-ICU and post-ICU treatment. Together with the medical expert, we found out, that decreased calcium levels (Hypocalcemia) are expected in ICU patients [3], which explains that development. Other attributes not being shown in Fig. 2 were also discussed, such as the glucose value, which did not make much sense, as it tends to change frequently. These attributes require a more fine granular process to make sense for observation through a process. However, the event attributes mentioned above do not tend to change frequently and can be compared department wise.

Another observation was, that patients visiting surgical departments have a stronger tendency to value changes in anion gab, bicarbonate, and calcium, represented by a higher RBC value, which could also be confirmed by the medical expert.

This presentation shows, that *dynamic* event attribute changes with their direction of change can be identified, allowing to derive additional insights out of data stored in event logs.

6   Discussion

This paper proposes an approach to detect changes in *dynamic* event attributes through the process by applying statistical tests on event attributes, relations, and trace variants. With that, we provide a method to analyse the behaviour of *dynamic* event attributes and allow identifying in which activities value changes occur.

We have shown an example use case in the healthcare domain and could confirm expected laboratory value behaviour, which was evaluated with a medical expert. As statistical tests are broadly accepted in the medical domain, it was possible to explain how we detect changes to the medical expert, who could understand the p-value and test statistics. We discussed, that a more fine granular process could bring additional insights, such as the comparison of different treatment paths and their laboratory value developments, allowing to evaluate, if different treatment activities have different effects on the patient's state.

However, we see potential for other application domains and do not want to limit the application to the healthcare domain. For example, other data intensive processes, such as manufacturing processes with sensor data, like temperature or vibration, could be of interest when looking at different manufacturing steps of one or multiple machines.

This contribution suggests identifying changes in three dimensions, which leads to a high amount of statistical tests conducted. Thus, we see one limitation in the exploration of changes, which is so far solved by looking at the statistical significant changes only from a two-dimensional perspective. Loops bring more complexity as well by adding more trace variants and relations when one is interested in comparing different loop iterations. Therefore, other perspectives or methods reducing cognitive load could be more suitable for different use cases. For example, when analysing loops, one could cluster the respective loop iterations according to their changing behaviour. The same holds for trace variants, which could be clustered as well.

Furthermore, the changes could be described in more detail by considering other aspects, such as time, resources, or other event attributes. For example, the longer one activity takes, the higher the difference between activities or the other way around. Additionally, changes in event attributes could be correlated with each other, such as creatinine and urea nitrogen in the evaluation.

The usage of statistical tests enables a detailed analysis of two samples, but requires a sufficient sample size as well. In general, the higher the sample size, the better the expressibility (power) of the test. Additionally, these tests cannot say that there is a guaranteed change for any given process instance, but can only give an indication that there is a non-random change in the given samples. Thus, there are almost always cases showing a changing behaviour and others do not. Understanding why some change and others do not is also not covered by us.

It should also be noted, that the statistical tests detect changes which go into one direction, such as from normal to high, resulting in a different distribution of the categories or continuous values. However, when we have changes in both directions, such as 50 from normal to high and 50 from high to normal, the marginal proportion is the same and no change would be detected. The same holds for continuous tests, where the RBC value would be close to 0 in this case. As the goal of this paper is to derive a common behaviour of *dynamic* event attributes in the process, this property suits us well. However, it might be interesting to investigate this kind of change and derive characteristics of increasing and decreasing cases.

In general, we see different use cases for change detection in *dynamic* event attributes. Besides exploring changes, one could also use this method to derive interesting variables for time-series machine learning tasks, such as process outcome prediction, by identifying process sensitive event attributes. Additionally, the changing behaviour could be used as a feature for decision mining, trace clustering or concept drift detection.

7   Conclusion and Future Work

This contribution researches methods to detect changes in *dynamic* event attributes from a three-dimensional perspective, represented as a change detection cube. This allows to understand the process behaviour of their actual values, as it can be seen between which process activities the values change. We see this method as a step forward to connect data-science with process science, allowing an even more comprehensible analysis of the data represented in event logs.

Future work could focus on enhancing the methodology by explaining the changes in more detail, for example, the correlation with other event attributes, such as time, or deriving characteristics of changing and non-changing cases. Additionally, other dimensions of change could be researched and evaluated regarding their suitability for different use cases. Lastly, the analysis of looping behaviour and trace variants could be improved by applying clustering, for example.

References

1. Process Mining. Springer, Heidelberg (2016). https://doi.org/10.1007/978-3-662-49851-4_16
2. Adams, J.N., van Zelst, S.J., Quack, L., Hausmann, K., van der Aalst, W., Rose, T.: A framework for explainable concept drift detection in process mining. In: Polyvyanyy, A., Wynn, M.T., Van Looy, A., Reichert, M. (eds.) BPM 2021. LNCS, vol. 12875, pp. 400–416. Springer, Cham (2021). https://doi.org/10.1007/978-3-030-85469-0_25
3. Afshinnia, F., Belanger, K., Palevsky, P.M., Young, E.W.: Effect of ionized serum calcium on outcomes in acute kidney injury needing renal replacement therapy: secondary analysis of the acute renal failure trial network study. Ren Fail **35**(10), 1310–1318 (2013)
4. Aminikhanghahi, S., Cook, D.J.: A survey of methods for time series change point detection. Knowl. Inf. Syst. **51**(2), 339–367 (2017)
5. Armstrong, R.A.: When to use the Bonferroni correction. Ophthal. Physiol. Opt. **34**(5), 502–508 (2014)
6. Bano, D., Zerbato, F., Weber, B., Weske, M.: Enhancing discovered process models with data object lifecycles. In: 2021 IEEE 25th International Enterprise Distributed Object Computing Conference (EDOC), pp. 124–133 (2021)

7. Berti, A., et al.: Process mining for python (pm4py): bridging the gap between process- and data science. CoRR abs/1905.06169 (2019). http://arxiv.org/abs/1905.06169

8. Chaudhuri, S., Dayal, U.: An overview of data warehousing and OLAP technology. SIGMOD Rec. **26**(1), 65–74 (1997)

9. Cooper, L.B., et al.: Serum bicarbonate in acute heart failure: relationship to treatment strategies and clinical outcomes. J. Card Fail **22**(9), 738–742 (2016)

10. Cremerius, J., Weske, M.: Data-enhanced process models in process mining (2021). https://arxiv.org/abs/2107.00565

11. Cremerius, J., Weske, M.: Supporting domain data selection in data-enhanced process models. In: Wirtschaftsinformatik 2022 Proceedings 3 (2022)

12. Ibanez-Sanchez, G., et al.: Toward value-based healthcare through interactive process mining in emergency rooms: the stroke case. Int. J. Environ. Res. Public Health **16**(10), 1783 (2019)

13. Jaber, S., et al.: Sodium bicarbonate therapy for patients with severe metabolic acidaemia in the intensive care unit (BICAR-ICU): a multicentre, open-label, randomised controlled, phase 3 trial. Lancet **392**(10141), 31–40 (2018)

14. Johnson, A., Bulgarelli, L., Pollard, T., Horng, S., Celi, L.A., Mark, R.: Mimic-iv (2020). https://doi.org/10.13026/A3WN-HQ05

15. Kerby, D.S.: The simple difference formula: An approach to teaching nonparametric correlation. Compr. Psychol. **3**, 11.IT.3.1 (2014)

16. de Leoni, M., van der Aalst, W.: Data-aware process mining: discovering decisions in processes using alignments. In: Proceedings of the 28th Annual ACM Symposium on Applied Computing, p. 1454–1461. SAC 2013. Association for Computing Machinery, New York, NY, USA (2013)

17. de Leoni, M., van der Aalst, W., Dees, M.: A general process mining framework for correlating, predicting and clustering dynamic behavior based on event logs. Inf. Syst. **56**, 235–257 (2016)

18. Mannhardt, F., de Leoni, M., Reijers, H.: The multi-perspective process explorer. In: CEUR Workshop Proceedings, vol. 1418, August 2015

19. Nguyen, H., et al.: Multi-perspective comparison of business process variants based on event logs. In: Conceptual Modeling, pp. 449–459. Springer International Publishing, Cham (2018). https://doi.org/10.1007/978-3-030-00847-5_32

20. Parab, S., Bhalerao, S.: Choosing statistical test. Int. J. Ayurveda Res. **1**(3), 187–191 (2010)

21. Reed, J.F., Salen, P., Bagher, P.: Methodological and statistical techniques: what do residents really need to know about statistics? J. Med. Syst. **27**(3), 233–238 (2003)

22. Schiffl, H.: Discontinuation of renal replacement therapy in critically ill patients with severe acute kidney injury: predictive factors of renal function recovery. Int. Urol. Nephrol. **50**(10), 1845–1851 (2018). https://doi.org/10.1007/s11255-018-1947-1

23. Sun, X., Yang, Z.: Generalized McNemar's test for homogeneity of the marginal distributions. In: SAS Global Forum, vol. 382, pp. 1–10 (2008)

Dealing with Unexpected Runtime Outcomes Within Process Models

Silvano Colombo Tosatto(✉) [ORCID] and Nick van Beest [ORCID]

DATA61, CSIRO, Sydney, Australia
{silvano.colombotosatto,nick.vanbeest}@data61.csiro.au

Abstract. Process models are designed to describe the required tasks to achieve a desired business goal. These models can be verified to be compliant with additional requirements, like regulations and business requirements. This means that process models can be designed and verified to behave according to some desired requirements. However, it is possible that some of the outcomes at runtime deviate from the design predictions of the model, which would render the model and the compliance verification obsolete. In this paper, we propose an approach aiming at detecting such runtime deviations through representing the tasks' outcomes as data ranges. When a deviation is detected, the approach re-evaluates compliance of the model given the unexpected outcomes during the execution, and if necessary and possible it adapts the remainder of the model's execution to pre-emptively avoid breaching the requirements.

Keywords: Business process compliance · Data ranges · Decidability · Runtime monitoring

1 Introduction

Former United States of America president Dwight D. Eisenhower once said *"In preparing for battle I always found that plans are useless but planning is essential"*. Part of the reason behind this quote is that rarely things go as planned, hence as they were designed they are often useless, but they can still be adapted to the current situation.

Process models are often used to describe plans, in other words the required steps to achieve a desired goal. Properties of these models have been studied in the past, such as for instance soundness [1], and compliance with respect to some given requirements [12]. Adapting to unpredicted situations has also received some attention: Maggi et al. [14] propose an approach based on DECLARE [17] and finite-state automata to identify unexpected scenarios where violations occur and if possible adapting them to address such violations.

In this paper, we study the problem of identifying unexpected behaviours and addressing the associated possible violations. We categorise unexpected behaviours of a model as the unexpected outcome of executing a task. Data is commonly used to describe the outcomes of executing a task, and we can identify unexpected outcomes when such do not align with the initial description. However when data is involved, decidability becomes an issue that needs consideration too, as shown by Bagheri Hariri

© Springer Nature Switzerland AG 2022
C. Di Ciccio et al. (Eds.): BPM 2022, LNBIP 458, pp. 173–189, 2022.
https://doi.org/10.1007/978-3-031-16171-1_11

et al. [4]. When the variables using to describe the outcomes of executing the tasks have an infinite domain of possible values that can be assigned to them, then the number of possible executions of the process models requiring to be checked also becomes infinite, leading the problem to become undecidable.

In this paper, we address the decidability issue by restricting the domains and collapsing the representation of these variables into ranges. Additionally, we show how process models decided to be compliant at design time can be adjusted when at runtime deviations over the expected behaviour are detected, leading to requirement violations.

The remainder of this paper is structured as follows. In Sect. 2, we position our work against existing research in the area of compliance and process adaptation. Next, Sect. 3 provides a formal background to process models and the regulatory framework for compliance. Subsequently, Sect. 4 describes the notion of full compliance in detail, while Sect. 5 formally defines runtime detection of deviations. Section 6 presents our method for repair of potential deviations to prevent compliance violations. Finally, we summarise our work in Sect. 7.

2 Related Work

Process models are designed so that they describe the allowed and required behaviour to achieve a business goal, while fulfilling all internal requirements (e.g. design properties, company policies) and external requirements (e.g. regulations). As such, these process models are typically ensured to be compliant by design, using various design-time compliance techniques [10, 19]. However, even assuming the actual behaviour of a process perfectly follows the normative process model, design-time correctness can only be guaranteed to a certain extent, as it cannot take into account runtime data. Although some approaches exist that take conditions into account and prove design-time compliance under those conditions [9], the actual behaviour is only known upon execution. Therefore, in most cases design-time verification should be considered a preventative measure that attempts to mitigate the risk of violating the requirements, as real-life execution may encounter deviations from the design.

Conformance checking is a procedure comparing executions as recorded in event logs with the desired behaviour as specified in the underlying process models to identify such deviations [2]. Regulations could be partially specified as models, allowing conformance techniques to identify mismatches with the actual executions. Nevertheless, these approaches are again predominantly control-flow [2, 6, 7]. Some approaches have been developed to verify event logs on their data as well. This after-the-fact analysis to prove compliance including the data perspective is referred to as *auditing*. However, none of these approaches have the ability to dynamically adapt running instances to either correct or prevent potential violations.

This is distinctly different from *model repair*, where the model is altered to more accurately reflect the actual behaviour as observed in an event log, usually by allowing inserting or skipping of activities (see e.g. [3, 18]). As such, the approach tries to optimise the model in terms of fitness, i.e. minimising the fraction of behaviour that is in the log but not possible according to the model. Although this allows to adapt processes based on insights from execution data (see e.g. [3], this is still a design-time approach that predominantly focuses on control-flow based adaptations and does not prevent a violation of requirements for running instances.

To facilitate runtime adaptation and repair, techniques from the field of automated planning have been integrated to allow dynamic on-the-fly reconfiguration of instances (see e.g. [5, 14, 16]). In these techniques, the domain is prespecified by means of actions (or tasks), which each have a set of corresponding preconditions (i.e. conditions that need to be satisfied prior to execution) and effects (representing the changes to the environment after execution). Using a predicate representing the goal to be achieved, automated planning aims to find a sequence of actions that achieve that goal starting from a given initial state. This approach has been applied in the field of business process management for the purpose of runtime process adaptation, where an external disruption or change in the data would cause the process to no longer be able to achieve its goal when following the process model [5]. As such, it is deployed to adapt the running process instance to still successfully achieve the intended goal where possible. Planning approaches require a fully specified domain, including pre-specifying every task that is potentially required to 'fix' the issue and may not be applicable in situations where information completeness can not be guaranteed [15, 16]. As such, it cannot provide suggestions on the data and its effects on future states. Additionally, the computational complexity on large data domains is typically high, resulting in a limited applicability in complex environments.

The preventative runtime approach presented in this paper uses data ranges to reduce the risk of non-compliance, while at the same time addressing the decidability issue.

3 Background

In this paper, we consider a simplified view of the process models we are evaluating. We consider process models as the set of their possible executions as illustrated by Definition 1, as the approach we propose in this paper to detect and address runtime deviations of the models does not need to consider the process' structure, but it focuses on its executions.[1]

Definition 1 (Process Model). *Given a process model P, we refer to the set of its executions as $\Sigma(P)$, where each execution ε is represented as an ordered sequence of tasks as follows: $\varepsilon = (t_1, t_2, \ldots, t_n)$.*

When proving whether a process model is *fully compliant* with some given regulatory requirements, each execution of the model must satisfy the requirements, represented as obligations, as described in Definition 2.

Definition 2 (Full Compliance). *Given a process P and a set of regulatory requirements represented as a set of obligations $\mathbb{O}$, P is fully compliant with respect to $\mathbb{O}$ if and only if every execution of P satisfies each obligation in $\mathbb{O}$.*

---

[1] We intentionally keep the definition of the base components of the problem abstract, as this allows the paper to focus on discussing the core issue: detecting and repairing runtime deviations in process models. As an additional bonus, using abstract descriptions to define the base components gives the proposed approach the flexibility to be applied to components fitting the features required by these abstract descriptions.

3.1 Satisfying Obligations

We use obligations to represent the regulatory requirements[2] that a process model must satisfy in order to be fully compliant. In particular, we use *achievement* and *maintenance* obligations, and we describe how these are satisfied in Definition 3.

Definition 3 (Achievement and Maintenance Obligations).
An obligation, defined as $\mathcal{O}^t\langle \pi, \tau, \delta\rangle$, is represented by the following elements:

- *A condition τ and a deadline δ, determining the* in force intervals *of an execution where the obligation's requirement (π) has to be checked. An* in force interval *is identified in an execution between states satisfying the condition (τ) and states satisfying the deadline (δ).*
- *A requirement π describing the satisfying condition that needs to be satisfied at least once within an* in force interval *when the type (t) of the obligation is an* achievement *(written as $\mathcal{O}^a$), or that needs to be satisfied in each state when the type is* maintenance *(written as $\mathcal{O}^m$).*

3.2 Executions and States

The elements of the obligations are evaluated along an execution to determine whether the obligation is satisfied by that particular execution. In particular, these elements are evaluated over the sequence of states describing an execution, where each state describes the situation holding after the execution of one of the tasks representing the execution. As an execution is a sequence of tasks, each task is associated with a state holding after its execution. Consequently, we can also consider an execution as a sequence of states.

The state of an execution after the execution of one of its tasks depends of the previous state and on the effects of the executed task. The effects of executing a tasks can be Boolean propositions, as well as variables. These are added to the previous execution's state when the task is executed. We use expected ranges to represent the possible values associated to a variable that can be obtained by the execution of a task, as shown in Definition 4. A similar approach has been adopted earlier by Knuplesch et al. [13] to deal with large domains.

Definition 4 (Variables and Execution's States). *The state of an execution can contain the following types of variables:*

propositional *the value of such variables can be either true or false and are represented by their value in the state, and the effects of the tasks, which updates the value contained in the execution's state.*
numerical *the value of these variables can be a number within a given interval. We represent this interval as $[a,b]$ in both the state and effects of the task's execution, where a is the lower bound and b is the upper bound. In general, we can consider*

---

[2] Note that obligations are not limited to represent and model regulatory requirements. For instance, an organisation could model the production requirements that a process must fulfil in each of its executions as obligations.

that the effects of executing a task alters the values in the state by increasing or decreasing them according to value assumed by the task's execution, and which is specified in the process specification. More complex interaction between effects and state variables can also be described in the process specification.

The advantage of using expected ranges is that we do not need to represent a different execution for each of the possible value, and the execution's states keep track of the values of the variables as intervals of possible values. This allows us to represent a single sequence of states for each possible sequence of tasks instead of having to explicitly consider the possible individual instantiations of the variables, and the combinatorial possible different sequences of states, which would quickly lead to the analysis of the possible executions to be intractable.

Example 1: Truck's Delivery Journey
Let us consider an example where a process model represents the possible delivery journeys of a truck. The process model shown in Fig. 1 represents these journeys, where a complete execution of the model represents a delivery journey of the truck, and a single task in the model represents a segment of the journey. The process model contains 4 possible executions: (t_1,t_2), (t_1,t_4), (t_3,t_2), and (t_3,t_4), corresponding to 4 possible delivery journeys.

An expected outcome is associated to each task. In this case, this represents the expected amount of hours required by the truck to travel the journey's segment represented by the task, and consists of a range included between a minimum and maximum value. When a task is executed (i.e. the truck travels the corresponding journey's segment) the amount of time is added to journey's time variable.

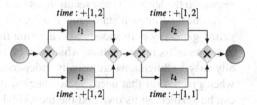

Fig. 1. Truck's delivery process model.

Finally, suppose that the truck is required to *complete its journey within 5 h*. We can verify whether this requirement is violated by checking the model, and in this case it is immediate to see that each execution of the model complies with the requirement, as the most time consuming routes $((t_1,t_2)$ and $(t_3,t_2))$ take between 2 and 4 h.

4   Full Compliance of Process Models

In this section, we describe how regulatory compliance of process models can be proven to be fully compliant with some given regulatory requirements expressed as obligations as discussed in Sect. 3.

To prove whether a process model is fully compliant, various techniques can be used, some focusing on analysing the structure of the models, like the approach proposed in [11], and others focusing on analysing the possible executions, like the Regorous approach [8] for instance. In this paper, we do not propose our own approach, but a protocol exploiting an existing compliance solution that focuses on proving full compliance of a process model through the analysis of its executions.

Here we outline the required properties for such a solution to be used in the proposed methodology to not only prove full compliance of a process model at design time, but also to both identify runtime deviations and if it is possible address them.

4.1 Compliance Checking: Required Properties

1. The procedure evaluates each execution of the process model independently.

   Discussion. A procedure evaluating the executions of the model has the advantage of disregarding the process' structure. That is, as long as a process model can generate a finite set of possible executions, these approaches are capable of evaluating the model by parsing the set. From a computational complexity perspective, the disadvantage of such approaches is that the amount of possible theoretical executions per model are either combinatorial with respect to the size of the model[3] or infinite when loops in the process models are unbounded.

2. When evaluating an execution, represented by the sequence of states constructed by the execution of the sequence of tasks, each obligation constituting the regulatory requirements is checked independently over the execution's states.

   Discussion. Verifying an obligation over an execution is linear in complexity with respect to the length of the execution, as the components of the obligations need to be checked in each state sequentially. Checking the components of an obligation generally corresponds to evaluating the truth value of a logical formula over one of its interpretation, which can be safely assumed to be computationally easy.[4] Obligations are usually independent from each other, meaning that when given a set that needs to be checked over an execution, each obligation can be verified on its own and the individual results can be easily aggregated.[5] Finally, evaluating the obligations individually also allows to pinpoint exactly where in the executions of the model the runtime violations occur. This additional information can be later used for further analysis aimed at addressing these regulatory issues, such as for instance adjusting the process model to avoid states where violations of the requirements occur.

---

[3] In practical scenarios it can be considered that the amount of possible executions of a model to be evaluated to be tractable. However, it must be taken into account that when heavy parallelisation is used, then the number of possible executions can become to large to be evaluated successfully.

[4] We would like to point out that it is still possible to have obligations whose components' verification is complex and can be associated to some more convoluted logics, however for the sake of clarity and simplicity we disregard these borderline cases in the present paper.

[5] Note that some more complex types of obligations may have their elements relate to the state of other obligations, such as for instance compensations, where the trigger of that kind of obligation corresponds to the violation of another. However, when these related obligations are organised in sequences and do not have circular relations, the verification procedure can still independently evaluate such sequences of related obligations and later aggregate the results.

An approach having the features mentioned above is sufficient to verify whether a process model is fully compliant with respect to a set of obligations representing the regulatory requirements. An approach having these properties is for instance Regorous [8], but other approaches having the same properties can be also used to determine compliance at design time. Furthermore we discuss in Sect. 5 how the same approaches can be reused to identify runtime deviations and determine whether they can be addressed.

5 Detecting Runtime Deviations

In Sect. 4 we have shown how a procedure with some specific properties can be used to prove that a process is fully compliant. A fully compliant process guarantees that each execution of a process model, whose task's execution conforms with the expected values represented by the *ranges*, is compliant with the regulatory requirements.

Given the problem specifications as described in Sect. 3, a *deviation*, a discrepancy from the expectations during an execution can be the following:

1. The order of the tasks being executed does not conform with the possible orders allowed by the structure of the process model.
2. The outcomes of completing one or more tasks during an execution of the model are not within the expected ranges specified at design time.

When either occur, the sequences of states resulting from the unexpected execution are not guaranteed to comply with the given regulatory requirements. In case of 1. the runtime execution wanders off from what is being expected by the process model, and the remainder of the execution cannot be predicted as it does not follow the process model's structure. Because of the unpredictability of this case, we focus on the second type of deviation, where the order of execution of the tasks is still one of the expected ones by the model's structure, but the outcome of executing a task resulted outside of the expected ranges. In the remainder of this section, we describe how to analyse the deviation to determine whether this can potentially lead to a violation. Additionally, in Sect. 6 we show for potential violations, how to analyse the deviation to determine whether and how the remainder of the execution can be adjusted to prevent these possible violations.

5.1 Detection

When a process model is being executed, its ongoing execution can be considered as a *partial execution*, corresponding to the prefix of one of the possible executions of the model. A *deviation* from the expected executions is defined and represented as described by Definition 5

Definition 5 (Deviation). *Given a process model P and one of its partial executions $\varepsilon_p = t_0, \ldots, t_n$, where t_n is the most recently executed task in ε_p, and we use $ev(t_n)$ to represent the variables' values representing the outcome of executing the task t_n.*

The execution of t_n in ε_p is a deviation if and only if:

- $ev(t_n) \notin ex(t_n)$, where $ex(t_n)$ are the set of ranges representing the expected outcomes from the execution of t_n as defined at design time of P.

We represent the deviation as $d(\varepsilon_p, \Delta)$, where Δ is the description of the effects discrepancies leading to $ev(t_n) \notin ex(t_n)$.

Identifying a *deviation* means that this particular execution is not among the ones that have been checked against the regulatory requirements during the compliance evaluation of the model. Thus, such execution cannot be guaranteed to be compliant, and further evaluations are required to determine that.

5.2 Assessment

After detecting a *deviation* during the execution of a process, it becomes desirable to determine how it impacts the ongoing execution. In other words it is desirable to determine whether the deviation can lead to violations of the regulatory requirements.

We can determine whether a *deviation* $(d(\varepsilon_p, \delta))$ can lead to violations in the continuation of the execution of the process (P) through the following steps:

1. Construct the *remainder process*[6] P' from P and $d(\varepsilon_p, \delta)$, such that the following property holds:
   - $\forall \varepsilon \in \Sigma(P)$ where ε_p is a prefix of ε, $\exists \varepsilon' \in \Sigma(P')$: $(\varepsilon_p + \varepsilon') \in \Sigma(P)$.
2. Check whether the *remainder process* (P') is fully compliant[7] with the regulatory requirements with the process' state holding after ε_p as the process starting state. If P' is not fully compliant, then there exists at least a possible future continuation of the execution ε_p that leads to a violation due to the deviation.

When checking the *remainder process* for full compliance using a procedure having the properties requested in Sect. 4.1, we know that each obligation representing the regulatory requirements is evaluated individually. This means that during the compliance checking process, when a violation is detected, it is immediate to identify the violated obligation. As the procedure evaluates the sequences of tasks and states representing the executions of the process model, the procedure also allows to identify the exact execution state where the violation occurs, which we can further analyse to determine the cause of the violation.

As checking compliance of a *remainder process* corresponds to analysing the predicted possible continuations of the execution where a deviation occurred, with further analysis we can try to determine whether it is possible to adjust the *outcome ranges* in the *remainder process* to avoid violating the given obligations. We describe this *reparation* procedure to prevent predicted possible violations in Sect. 6.

---

[6] Given the running partial execution ε_p of P, the associated *remainder process* P' represents with its possible executions the possible continuations of ε_p as determined in P.

[7] Any technique used to check compliance of the original process can be reused to verify compliance for the *remainder process*.

Example 2: Truck's Journey: Deviation

Let us consider a runtime execution of the truck's journey process model shown in Example 1. Let us assume that in this runtime execution, the first task being executed is t_1 and the returned outcome to be 3. That is, the truck took 3 h to travel this journey's segment, and the time taken was longer than the amount predicted at the design time of the process model. The execution of t_1 is classified as a *deviation*, and the *remainder model* (Fig. 2) is used to evaluate its effects.

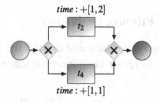

Fig. 2. Truck's remainder model.

A remainder model describes the possible continuations of an ongoing execution, in this case of the deviating execution where t_1 has executed and the deviation was detected. Because of the outcome of executing t_1, the starting state of the remainder model has its *journey_time* variable set to 3. Evaluating compliance of the remainder model against the requirement that the truck is required to *complete its journey within 5 h*, we can immediately see that executing t_2 leads to *journey_time* $= [4, 5]$. This means that the detected deviation can potentially lead to a violation of the requirement in case the amount of hours to travel in t_2 is 2 h.

We can try to avoid the predicted violation by adjusting the estimated outcomes in the remainder model. Reducing the expected outcome range of t_2 to $[1, 1]$ leads to the remainder model to fully comply with the requirement. That is, continuing the execution with the adjustment in mind would always be compliant with the requirements, unless further deviations occur, which would potentially require additional analysis.

6   Reparation

Considering a *remainder model* following from a *deviation*, which has shown to contain some violations in its executions, a *reparation* consists of modifying the ranged variables representing the outcomes of executing the tasks of the remainder model in such a way that no violations occur if possible.[8] If some of the violations cannot be prevented, then we consider the model, and in turn the deviation, to be *not repairable*.

6.1   Computational Complexity

The reparation problem entails finding a range assignment for each of the ranged variables in each of the tasks of the model in such a way that the resulting executions satisfy each of the obligations representing the regulatory requirements.

In theory, the possible ranges that can be assigned to the variables representing the outcomes of the tasks are unbounded, meaning that there are infinite ways in which these ranges can be assigned to the variables. This can lead to an infinite search space

---

[8] The reparation problem presents many similarities with the constraint satisfaction problems, which can be described as the problems of finding the values assignments of some variables such that the resulting state satisfies some given constraints.

for the problem's solution, which can be an issue when using a naive algorithm blindly looking into the search space for one fitting the constraints, as there are no guarantees that the algorithm would ever terminate.

In the remainder of this section, we focus on describing how to handle the vastness of the search space of the problem. First, we discuss some obligations' evaluation orders and preferences over alternative alterations options to reduce the search space required to be explored. Second, we introduce a procedure adopting these preferences and evaluation orders.

6.2 Effective Evaluation

We now introduce and discuss the individual features that our proposed practical approach to repair violations originating from runtime deviations.

Iterative Detection and Handling of Violations. The first and least surprising feature concerns how to detect which alterations are required to be performed over the model. The worst scenario in this case would be an approach picking a random solution from the search space and then evaluate whether the solution fits the requirements.

A better way to explore the search space instead of blindly looking at the possibilities, is to look at the solutions tied to solving some of the violations that affect the executions of the model. This is achieved by iterating over the obligations and for each of them determine whether the model violates them. For each detected violation we can then apply the changes required to address it. This effectively limits the exploration of the search space to those affecting at least some of the existing violations.

This way of iterating over the detected violations and altering the process to fix them presents some problems that need to be considered. When altering the process to fix a violation, there exists the possibility that the change introduces violations for other obligations. As a result, each obligation should be re-checked over the process after each change, and the potential for live-locks implies that in some cases the reparation procedure would not terminate (Example 3).

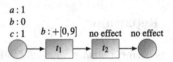

Fig. 3. Simple linear model.

Example 3: Live-Lock on Repair
Consider two obligations: $\mathcal{O}_1^m(a == 1, b < 10, \bot)$ and $\mathcal{O}_2^a(c == 1, b \geq 10, \bot)$, and the linear *remainder model* shown in Fig. 3, where the effects of executing a task are annotated above the task itself. Above Start, we annotate the starting state of the execution of the model, namely a and c having value 1, and b having value 0. For the sake of the example, let us assume that $c = 1$ in the starting state is the result of a *deviation*. The starting state of the *remainder model* sets in force both obligations $\mathcal{O}_1$ and $\mathcal{O}_2$. With only t_1 manipulating the value of b, we have that $\mathcal{O}_1$ is satisfied, as the value will be added up to at most 9, while $\mathcal{O}_2$ cannot be satisfied as the value will never go above 9.

Repairing the violation of $\mathcal{O}_2$ consists of setting the outcome of t_1 to $b : +[10,10]$, which allows to satisfy the requirement. However, it is immediate to see that the adjustment introduces a violation for $\mathcal{O}_1$. Fixing the violation of $\mathcal{O}_1$ would violate $\mathcal{O}_2$ again, causing the procedure to enter a live-lock loop.

Avoid Re-evaluation. When changes are introduced in a model, obligations need to be re-evaluated to determine whether the new configuration complies with them, even those that have already been evaluated. This is computationally expensive and can in the worst case scenario lead to live-locks as shown in Example 3.

To avoid re-evaluating obligations that have already been evaluated on the model, we constrain the allowed alterations to the ones that ensures that the compliance state of obligations already evaluated does not change. We achieve this by not allowing alterations that target variables that can affect the compliance state of obligations that have already been checked. In other words, when an obligation is checked and its violations repaired, the variables influencing the compliance state of that obligation cannot be altered by further reparations.

By preventing some variables to be altered, the search space is greatly reduced, which comes with both positive and negative consequences: on the positive side, any constraint reducing the explorable search space improves the performances of algorithms trying to find solutions. However, these constraints also run the risk of possibly hiding some of the solutions in spaces that are no longer explorable. To minimise the amount of non-explorable solutions, we use an evaluation order that minimise the constraints over further reparations.

Maximise Repair Options. To avoid re-evaluation we can consider that the variable influencing the compliance state of an obligation to be locked from further changes after that obligation has been evaluated and its violations repaired. These *locked* variables hinder the possibilities of repairing further obligations, as fewer options remain available, as some are not an option because of some locked variables.

For every pair of obligations being evaluated, to minimise the impact of locked variables, the obligation locking fewer variables for the other should be evaluated and repaired earlier. This means that the impact on the ability of repairing violations for the obligation evaluated later is lower. Between two obligations we can define the relatively least restricted as the one being influenced by the higher number of variables that are not influencing the other obligation (Definition 6 and Definition 7). The least restricted obligation should be verified later than the other, as it would retain more options to repair the eventual violations. Additionally, note that when the sets of variables influencing two obligations are completely disjoint, then these obligations are independent and their relative evaluation order is irrelevant.

Definition 6 (Related Variables). *Given an obligation $\mathcal{O}(r,t,d)$, let the function s return the ranged variables involved in the evaluation of the element of an obligation, such as for instance $s(r)$ returns the ranged variables involved in the evaluation of the trigger of $\mathcal{O}$. Furthermore, we define: $s(\mathcal{O}) = s(r) \cup s(t) \cup s(d)$.*

Definition 7 (Altering Relations). *Given two obligations* $\mathscr{O}_1(r,t,d)$ *and* $\mathscr{O}_2(r,t,d)$, $\mathscr{O}_1$ *is altering related to* $\mathscr{O}_2$ *if and only if:* $s(\mathscr{O}_1) \cap s(\mathscr{O}_2) \neq \emptyset$.

We use $sr(\mathscr{O}_1, \mathscr{O}_2) = s(\mathscr{O}_1) \cap s(\mathscr{O}_2)$ *to refer to the set of relation sets between two obligations.*

Finally, from Theorem 1 we know that we cannot have circular relations. Thus, considering the set of ordering relations between pairs of obligations as a partial ordered set, then there exists always at least a total order satisfying the partial order.

Theorem 1 (Non-circularity). *Given three sets* $A, B,$ *and* C, *the following set of properties is inconsistent:*

1. $|A \setminus B| > |B \setminus A|$
2. $|B \setminus C| > |C \setminus B|$
3. $|C \setminus A| > |A \setminus C|$

Proof.
We can rewrite the three inequalities using the labels shown in Fig. 4 to refer to the various parts of the intersections between the sets A, B, and C as follows:

1. $|a \cup e| > |b \cup f|$
2. $|b \cup d| > |c \cup e|$
3. $|c \cup f| > |a \cup d|$

As the sets labelled in Fig. 4 are disjoint, we can rewrite the system using numeric variables (maintaining the same label) referring directly to the cardinality of the intersections as follows:

(i) $a + e > b + f$
(ii) $b + d > c + e$
(iii) $c + f > a + d$

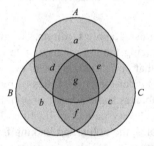

The obtained inequalities system is a contradiction:

Fig. 4. Set intersections.

1. **(i)** can be rewritten as $a > b + f - e$
2. **(ii)** can be rewritten as $b - e > c - d$
3. **(iii)** can be rewritten as $c + f - d > a$
4. From 1. and 3. and transitivity: $c + f - d > b + f - e$ which simplifies to $c - d > b - e$
5. 2. and 4. are *contradictory*.

Having shown that the system of inequalities has no solution, we have proven that Theorem 1 is correct. □

Minimise Change. When altering a set of ranged variables to prevent a violation in the model, the preferred set to be used is the one that produces the least, and the safest alterations. Fewer alterations are preferred because they maintain the behaviour closer to the original. For this reason, when multiple alteration options are available to prevent a violation in a model, the one involving the fewer variables to be altered is the one that should be preferred.[9]

We consider safer alterations as a secondary preference measure. Preventing the violation of an obligation can be done by dealing with either of its elements: the trigger, deadline, or requirement. We discuss in Definition 8 how each element can be influenced by altering some ranged variables and how the effects lead to preventing the violation.

Definition 8 (Repair Options).

There are three options to repair a violation. Considering that a violation of an obligation is related to one of the in force instances of the obligation itself, we list the three repair options below:

1. *Preventing the trigger: adjust the range of an outcome of a task's execution to prevent the satisfaction of the obligation's trigger. This adjustment prevents the instance of the obligation to come in force, in turn preventing its violation.*
2. *Fixing the requirement: adjust the range of an outcome of a task's execution to satisfy the obligation's requirement. For achievement obligations this would try to manipulate the execution's states to construct one satisfying the requirement before the deadline determining the violation. Differently for maintenance obligation, this option aims at preventing the state where the requirement fails.*
3. *Manipulating the deadline: adjust the range of an outcome of a task's execution to change the state satisfying the obligation's deadline. For achievement obligations this consists of changing the execution's states to postpone the state in which the deadline is satisfied, hence preventing the identified violation. For maintenance obligations this consists of anticipating the deadline to an earlier state, which allows to avoid the state in which the requirement failed to be maintained.*

We discuss in Definition 9 which of the elements should be preferred as the alteration target, and this additional preference order can be used as a secondary preference in addition to the cardinality of the change.

Definition 9 (Repair Choices Order). *Depending on the type of the obligation we provide and motivate an intuition based preference order in which the repair options should be applied to try to fix a violation.*

Achievement

1. *Fixing the requirement: this is the preferred method as it allows to successfully terminate the in force instance of the obligation and avoid the violation at the state where the deadline occurs.*

---

[9] For simplicity, we consider as the measure of the alteration only the cardinality of the set of variables altered, and not the magnitude. However, both measures should be considered when deciding which alteration is the least impacting on the current behaviour, and we plan to investigate this in our future research.

2. *Preventing the trigger: this is the second preferred method as it prevents the coming in force of the instance. This is less preferable than* fixing the requirement *as preventing the trigger may be actually just delaying it, hence not avoiding the violation in the same execution state, and requiring another* repair *round.*
3. *Manipulating the deadline: the least preferred option for achievement obligations, as this simply delays the evaluation of the in force instance, keeping it in an unknown state and potentially requiring later along the execution to* repair *another violation related to the same in force instance.*

Maintenance

1. *Manipulating the deadline: for maintenance obligations this is the preferred method, as it would allow to terminate the in force instance successfully in an earlier state than the one where the identified violation occurs.*
2. *Preventing the trigger: this is not the most preferable choice for the same reason as for achievement obligations.*
3. *Fixing the requirement: the least preferred option as it brings the in force instance into an unknown state, and leaving it vulnerable to future violations along the continuation of the execution.*

6.3   A Practical Approach

We proceed with the algorithm to identify violations given a deviation, and subsequently repair them following the principles discussed in Sect. 6.2.

Algorithm 1 (Identify). *Given a* remainder process *P, and a set of obligations $\mathbb{O}$:*

1. *Compute the* Altering Relations *(Definition 7) partial order of $\mathbb{O}$*
2. *Sort $\mathbb{O}$ according to the partial orders indentified*
3. *For each obligation $\mathcal{O}$ in $\mathbb{O}$*
   (a) *Evaluate compliance$(P, \mathcal{O})$[10]*
       i. *If a violation v is detected:*
           A. *Repair the* remainder process*: $repair(P, v)$*
           B. *If the result of repair is* unrepairable*, then terminate with failure,*
           C. *otherwise return to 3.(a) to evaluate $\mathcal{O}$ over the repaired* remainder process
       ii. *Otherwise add the variables influencing $\mathcal{O}$ to the set of locked variables $(\mathbb{LV})$ and proceed with the next obligation in $\mathbb{O}$*
4. *Terminate with successful repair of each violation introduced by the deviation*

Algorithm 1 evaluates each obligation over the *remainder model*, and for each violation a reparation is performed. Note that when a violation is identified and repaired, the obligation is re-evaluated to determine if additional violations exist. As the *remainder model* is repaired as soon a violation is detected for an obligation, the same obligation needs to be evaluated again to verify that no further violations exist. The algorithm

---

[10] Note that this does not refer to a particular compliance checking approach, but any approach satisfying the properties described in Sect. 4.1 can be used.

either terminates with a failure when a violation cannot be repaired, or successfully after having verified that each obligation is not violated by the repaired *remainder model*.

The time complexity of Algorithm 1 is mainly governed by the size of the obligations being checked (3.) and the complexity of the compliance checking algorithm being applied (3.a). Assuming the latter to be linear with respect to the executions of the model leads the algorithm to require $|\mathbb{O}| \times |\Sigma(P)|$ iterations. We must also consider the complexity of *repair* (Algorithm 2), which in the worst case corresponds to a number of iterations equal to the length of the execution being evaluated. Leading to the updated time complexity: $|\mathbb{O}| \times |\Sigma(P)| \times |\varepsilon|$ where $|\varepsilon|$ is the maximal length of an execution belonging to P.

Algorithm 2 (Repair). *Given a set of locked variables* $\mathbb{LV}$, *a remainder process* P, *and a violation* v, *where* v *contains the following data:*

- *The obligation* $\mathcal{O}$ *violated*
- *The execution* ε *identifying the violation*

1. *Identify the repair options* $\mathbb{RO}$ *from* $options(\varepsilon, v)$[11]
2. *Remove from* $\mathbb{RO}$ *each option containing variables in* $\mathbb{LV}$
3. *If* $\mathbb{RO}$ *is empty, terminate with* unrepairable
4. *Otherwise:*
    (a) *Select the minimal*[12] *repair option* ro *in* $\mathbb{RO}$
    (b) *Apply the repair* ro *to* P *and return the repaired process for further evaluation*

Algorithm 2 determines the possible options involving non-locked variables to repair the detected violation, and the option minimising change is selected as the repair candidate. When no repair options are available, the process is determined to be unrepairable given the detected violation for the deviation being investigated. Determining the repair options involve investigating the available ones, and excluding the ones that cannot be used due to locked variables. The computational complexity of determining this set is related to the size of the possible options: $|\mathbb{RO}|$.

The proposed approach limits the search space of the problem thanks to a combination of locking variables from being changed further, and specifying an order in which the obligations are evaluated. However, the current limits may exclude some genuine solutions to be considered, and in the worst case scenario[13] it could happen that the only existing solutions are excluded by the limitations over the search space. We plan to investigate refinements for this restriction when extending this approach.

---

[11] We do not provide a detailed definition for the procedure identifying the possible repair options given a violation resulting from a behavioural deviation. While this is definitely an interesting problem that we plan to tackle in our future research, it can be considered an orthogonal problem and to minimise the scope of the paper we keep the focus on the main procedure, while assuming this auxiliary procedure as given for now.

[12] That is, we consider the smallest cardinality and when multiple options are still available, the order used in Definition 9 is used to further reduce the amount of candidates.

[13] This can be the case where deviations during an execution leads to a large amount of failures for many obligations governing the model. More general cases would be represented by small deviations leading to a few violations that can be then resolved by iterating the approach a limited number of times.

7 Summary

In this paper, we investigated the problem of detecting runtime deviations of process models and dealing with the eventual resulting violations. We discussed the major computational issues when dealing with such problem, namely the vast search space in which algorithms need to look for a solution. We also outlined some procedures that exploit properties of the problem to reduce this search space, in particular preventing the risk of live-locks by evaluating the required obligations in a specific order and avoiding to have to re-evaluate them.

Our approach relies on *full compliance*, which means that executing a model with such property would always satisfy the specified requirements. While requiring full compliance is generally more restrictive than partial compliance, this allows us to treat these models as *fire and forget* entities that can be executed without further supervision from the user, while the proposed repair approach allows these models to adapt to some unforeseen scenarios automatically.

To the best of our knowledge this is a first approach dealing with the problem of adapting process models to detected runtime deviations during their execution, and while the currently proposed approach is quite coarse, it outlines the relevant features of the problems, which we plan to tackle again in our future research to provide a finer and more precise solution.

References

1. Aalst, W.. M.. P..: Verification of workflow nets. In: Azéma, Pierre, Balbo, Gianfranco (eds.) ICATPN 1997. LNCS, vol. 1248, pp. 407–426. Springer, Heidelberg (1997). https://doi.org/10.1007/3-540-63139-9_48
2. van der Aalst, W.M.P., Adriansyah, A., van Dongen, B.: Replaying history on process models for conformance checking and performance analysis. Wiley Interdisciplinary Rev. Data Mining Knowl. Discovery 2(2), 182–192 (2012)
3. Armas Cervantes, Abel, van Beest, Nick R. T. P.., La Rosa, Marcello, Dumas, Marlon, García-Bañuelos, Luciano: Interactive and incremental business process model repair. In: Panetto, Hervé, Debruyne, Christophe, Gaaloul, Walid, Papazoglou, Mike, Paschke, Adrian, Ardagna, Claudio Agostino, Meersman, Robert (eds.) OTM 2017. LNCS, vol. 10573, pp. 53–74. Springer, Cham (2017). https://doi.org/10.1007/978-3-319-69462-7_5
4. Bagheri Hariri, B., Calvanese, D., De Giacomo, G., Deutsch, A., Montali, M.: Verification of relational data-centric dynamic systems with external services. In: Proceedings of the 32nd ACM SIGMOD-SIGACT-SIGAI Symposium on Principles of Database Systems, pp. 163–174. Association for Computing Machinery, New York (2013)
5. van Beest, N.R.T.P., Kaldeli, E., Bulanov, P., Wortmann, J.C., Lazovik, A.: Automated runtime repair of business processes. Inf. Syst. 39, 45–79 (2014)
6. Carmona, J., van Dongen, B.F., Solti, A., Weidlich, M.: Conformance Checking-Relating Processes and Models. Springer International Publishing (2018)
7. García-Bañuelos, L., Van Beest, N.R.T.P., Dumas, M., La Rosa, M., Mertens, W.: Complete and interpretable conformance checking of business processes. IEEE Trans. Software Eng. 44(3), 262–290 (2017)
8. Governatori, G.: A short introduction to the regorous compliance by design methodology. In: Proceedings of the 1st Workshop on Technologies for Regulatory Compliance co-located with the 30th International Conference on Legal Knowledge and Information Systems (JURIX). CEUR Workshop Proceedings, vol. 2049, pp. 7–13 (2017)

9. Groefsema, H., van Beest, N.R.T.P., Armas-Cervantes, A.: Efficient conditional compliance checking of business process models. Comput. Ind. **115**, 103181 (2020)

10. Groefsema, H., van Beest, N.R.T.P., Aiello, M.: A formal model for compliance verification of service compositions. IEEE Trans. Serv. Comput. **11**(3), 466–479 (2018)

11. Groefsema, H., van Beest, N.R., Aiello, M.: A formal model for compliance verification of service compositions. IEEE Trans. Serv. Comput. **11**(3), 466–479 (2016)

12. Hashmi, M., Governatori, G., Lam, H.-P., Wynn, M.T.: Are we done with business process compliance: state of the art and challenges ahead. Knowl. Inf. Syst. **57**(1), 79–133 (2018). https://doi.org/10.1007/s10115-017-1142-1

13. Knuplesch, D., Ly, L.T., Rinderle-Ma, S., Pfeifer, H., Dadam, P.: On enabling data-aware compliance checking of business process models. In: Parsons, J., Saeki, M., Shoval, P., Woo, C., Wand, Y. (eds.) ER 2010. LNCS, vol. 6412, pp. 332–346. Springer, Heidelberg (2010). https://doi.org/10.1007/978-3-642-16373-9_24

14. Maggi, Fabrizio Maria, Marrella, Andrea, Capezzuto, Giuseppe, Cervantes, Abel Armas: Explaining non-compliance of business process models through automated planning. In: Pahl, Claus, Vukovic, Maja, Yin, Jianwei, Yu, Qi. (eds.) ICSOC 2018. LNCS, vol. 11236, pp. 181–197. Springer, Cham (2018). https://doi.org/10.1007/978-3-030-03596-9_12

15. Marrella, Andrea: What automated planning can do for business process management. In: Teniente, Ernest, Weidlich, Matthias (eds.) What automated planning can do for business process management. LNBIP, vol. 308, pp. 7–19. Springer, Cham (2018). https://doi.org/10.1007/978-3-319-74030-0_1

16. Marrella, A.: Automated planning for business process management. J. Data Semantics **8**(2), 79–98 (2019)

17. Pesic, M., Schonenberg, H., van der Aalst, W.M.P.: Declare: full support for loosely-structured processes. In: IEEE EDOC, pp. 287–287 (2007)

18. Polyvyanyy, A., van der Aalst, W.M.P., ter Hofstede, A.H.M., Wynn, M.T.: Impact-driven process model repair. ACM Trans. Softw. Eng. Methodol. **25**(4), 1–60 (2016)

19. Sadiq, Shazia, Governatori, Guido, Namiri, Kioumars: Modeling control objectives for business process compliance. In: Alonso, Gustavo, Dadam, Peter, Rosemann, Michael (eds.) BPM 2007. LNCS, vol. 4714, pp. 149–164. Springer, Heidelberg (2007). https://doi.org/10.1007/978-3-540-75183-0_12

Detecting Context-Aware Deviations in Process Executions

Gyunam Park[✉], Janik-Vasily Benzin, and Wil M. P. van der Aalst

Process and Data Science Group (PADS), RWTH Aachen University,
Aachen, Germany
{gnpark,wvdaalst}@pads.rwth-aachen.de, janik.benzin@rwth-aachen.de

Abstract. A deviation detection aims to detect deviating process instances, e.g., patients in the healthcare process and products in the manufacturing process. A business process of an organization is executed in various contextual situations, e.g., a COVID-19 pandemic in the case of hospitals and a lack of semiconductor chip shortage in the case of automobile companies. Thus, *context-aware deviation detection* is essential to provide relevant insights. However, existing work 1) does not provide a systematic way of incorporating various contexts, 2) is tailored to a specific approach without using an extensive pool of existing deviation detection techniques, and 3) does not distinguish *positive* and *negative* contexts that justify and refute deviation, respectively. In this work, we provide a framework to bridge the aforementioned gaps. We have implemented the proposed framework as a web service that can be extended to various contexts and deviation detection methods. We have evaluated the effectiveness of the proposed framework by conducting experiments using 255 different contextual scenarios.

Keywords: Context-aware deviation detection · Context · Deviation detection · Process mining

1 Introduction

Deviation detection in process executions aims to identify anomalous executions by distinguishing deviating behaviors from normal behaviors. A range of deviation detection techniques for business processes has been proposed [4]. The techniques are categorized as supervised and unsupervised ones. The former defines normal behavior to identify deviations of recorded process executions with respect to the specified normal behavior, whereas the latter identifies deviations without such normal behaviors. Since many businesses lack the specification of normal behavior, unsupervised deviation detection techniques recently gained more attention [4].

As a process is executed in a specific *context* (e.g., COVID-19 Pandemic) that affects the behavior of the execution, it is indispensable to consider the context when detecting deviations [2]. In this regard, *context-aware deviation detection* aims to classify a trace (i.e., a sequence of events by a process instance) to ①

© Springer Nature Switzerland AG 2022
C. Di Ciccio et al. (Eds.): BPM 2022, LNBIP 458, pp. 190–206, 2022.
https://doi.org/10.1007/978-3-031-16171-1_12

context-insensitive normal meaning the trace is normal regardless of context, ②
context-insensitive deviating meaning the trace is deviating regardless of context,
③ *context-sensitive normal* meaning the trace is deviating without considering
context but normal when considering context, and ④ *context-sensitive deviating*
meaning the trace is normal without considering context but deviating when
considering context.

Few approaches have been developed to (indirectly) solve the context-aware
deviation detection problem [4]. For instance, Pauwels et al. [15] extend Bayesian
networks to learn conditional probabilities for organizational contexts such as
roles of resources. Warrender et al. [17] propose a sliding-window based approach
that considers time-related context. Mannhardt et al. [12] conceptualize context
as data attributes of process instances.

However, each approach is tailored to consider limited aspects of contexts, not
providing a systematic way to extend the approach to consider various aspects of
contexts. Given a large space of possibly relevant contexts proposed in studies on
contexts (cf. Subsect. 2.2), we need a systematic framework to integrate context
to deviation detection.

Moreover, a framework to integrate a large number of existing deviation
detection methods with different strengths and weaknesses on varying assumptions is missing. Instead, the existing work is confined to a single method and
inherits the methods' unique set of properties.

Furthermore, existing techniques do not distinguish *positive* and *negative* contexts. The former justifies deviations. For instance, COVID-19 Pandemic in a
healthcare process explains the long waiting time for admission, e.g., due to the
sudden increase in the number of patients. The latter refutes non-deviations.
"Crunch time" in a video game industry denies a normal throughput time of
the game development process, e.g., with the compulsory overwork by employees.
Existing work considers only negative contexts when integrating context into deviation detection.

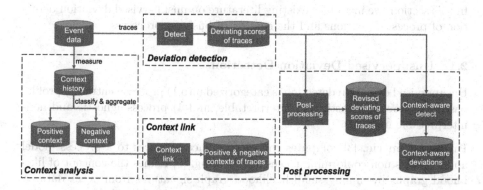

Fig. 1. An overview of the framework for context-aware deviation detection

In this paper, we propose a framework based on *post-processing mechanism* to systematically support the context-aware deviation by integrating the extensive existing deviation detection methods and contexts. As shown in Fig. 1, the framework consists of four components. First, *deviation detection* computes deviating scores of traces, with which we can classify *non-context deviating* and *non-context normal* traces. Next, *context analysis* computes *positive* and *negative* contexts by aggregating *context history*. Afterwards, *context link* connects the context to traces. Next, *post-processing* increases the deviation score of a trace with the positive context of the trace and decreases it with the negative context. Using the revised deviation score, we classify traces as *context-normal* and *context-deviating*. Finally, we label a trace as one of ①-④ based on the *non-context* and *context* classifications.

To summarize, this paper provides the following contributions:

- We propose a framework to solve the context-aware deviation detection problem while integrating the existing deviation detection methods and contexts.
- We extend the context conceptualization with *positive* and *negative* contexts that carry dedicated semantics for deviation detection.
- We implement a flexible and scalable web service supporting the framework and evaluate the effectiveness of the framework with 225 simulated scenarios.

The remainder is organized as follows. We discuss the related work in Sect. 2. Then, we present the preliminaries in Sect. 3. Next, we introduce the context-awareness in Sect. 4 and a framework for integrating contexts and deviation detection in Sect. 5. Afterward, Sect. 6 introduces the implementation of a web application, and Sect. 7 evaluates the effectiveness of the proposed framework. Finally, Sect. 8 concludes the paper.

2   Related Work

In this section, we introduce existing literature on unsupervised deviation detection of process executions and the context of business processes.

2.1   Unsupervised Deviation Detection

Unsupervised deviation detection is categorized into 1) process-centric, 2) profile-based, 3) process-agnostic and interpretable, and 4) process-agnostic and non-interpretable methods.

Process-Centric. [3] computes the conformance of traces to a process model and classifies non-conforming traces as deviating. [5] refines the concept of likelihood graphs by mining small likelihood graph signatures from event data. A deviation is determined by comparing the execution likelihood of a trace with respect to a set of mined signatures and a reference likelihood. [8] discovers process models using genetic algorithms and conducts conformance checking using token-based replay to detect deviating traces.

Profile-Based. [11] iteratively samples more normal sets of traces and *profiles* each trace against the more normal set of traces. The result is a sorted list of traces according to their profiles in the last iteration, which is used to partition the event data into a set of normal traces and a set of deviating traces using a deviation threshold.

Process-Agnostic and Interpretable. [15] extends Bayesian networks and defines a conditional likelihood-based score using the extended bayesian network on traces. All traces are then sorted according to the score, and the first k are returned as deviating traces. [6] uses *association rules*. A set of anomaly detection association rules specifying normal behavior is mined from the event data. A trace is detected as deviating if its aggregate support is below the aggregate support of its most similar trace in the event data with respect to the set of anomaly detection association rules. [17] uses a sliding window-based approach to extract frequency information over those windows. If a trace contains infrequent windows, then it is deviating.

Process-Agnostic and Non-interpretable. [13] encodes traces in event data using one-hot encoding and train *autoencoder neural network* with them. The deviation of a trace is determined using the error the autoencoder makes in predicting the trace. [14] further develops the application of neural networks to event data by training a recurrent neural network to predict the next event in integer-encoding based on the current event in a trace. The aggregate likelihood of predicting the correct events is used to detect deviations.

Some of the unsupervised deviation detection methods provide room for handling limited kinds of context but take method-dependent approaches such that neither a general integration nor support for a systematic extension of context is provided. In this work, we provide a general framework to integrate various unsupervised deviation detection techniques with different strengths, weaknesses, and assumptions to systematically extend them with contexts.

2.2 Context

In pervasive computing, especially for developing adaptive services, context is conceptualized as the lower level of the abstraction of raw data [18]. Another higher-level abstraction, called situation, is introduced to map one or multiple contexts to semantically richer concepts such as users' behaviors.

In business processes, a context is a multitude of concepts that affect the behavior and performance of the process. [2] derives four levels of context that should be considered during the analysis of processes to improve the quality of results. [16] extends it and provides an ontology of contexts in BPM by conducting an extensive literature review of the context in BPM.

More ontological approaches have been proposed to specify context and situations. Generally, they categorize contexts into *intrinsic* and *relational*. [7] differentiate between *intrinsic* and *relational* context whereby intrinsic context is essential to the nature of the entity and relational context is inherent to the relation of multiple entities. [9] develops a two-level framework for structuring context, which is more coarse-grained than the four levels of [2].

In this work, we merge relevant contexts of the earlier work and their categorizations into an integrative context ontology that is aimed at extracting context from event data.

3   Preliminaries

Definition 1 (Event). *Let $\mathbb{U}_e$ be the universe of events, Let $\mathbb{U}_{att}=\{act, case, time, \dots\}$ be the universe of attribute names. For any $e \in \mathbb{U}_e$ and $att \in \mathbb{U}_{att}$: $\#_{att}(e)$ is the value of attribute att for event e, e.g., $\#_{time}(e)$ indicates the timestamp of event e.*

Definition 2 (Trace). *A trace is a finite sequence of events $\sigma \in \mathbb{U}_e^*$ such that each event appears only once, i.e., $\forall_{1 \leq i < j \leq |\sigma|} \, \sigma(i) \neq \sigma(j)$. Given $\sigma \in \mathbb{U}_e^*$ and $e \in \mathbb{U}_e$, we write $e \in \sigma$ if and only if $\exists_{1 \leq i \leq |\sigma|} \, \sigma(i) = e$. We define $elem \in \mathbb{U}_e^* \to \mathcal{P}(\mathbb{U}_e)$ with $elem(\sigma) = \{e \in \sigma\}$.*

Definition 3 (Event Log). *An event log is a set of traces $L \subseteq \mathbb{U}_e^*$ such that each event appears at most once in the event log, i.e., for any $\sigma_1, \sigma_2 \in L$ such that $\sigma_1 \neq \sigma_2 : elem(\sigma_1) \cap elem(\sigma_2) = \emptyset$. Given $L \subseteq \mathbb{U}_e^*$, we denote $E(L) = \bigcup_{\sigma \in L} elem(\sigma)$.*

Definition 4 (Time Window). *Let $\mathbb{U}_{time}$ be the universe of timestamps. $\mathbb{U}_{tw} = \{(t_s, t_e \in \mathbb{U}_{time} \times \mathbb{U}_{time} \mid t_s \leq t_e\}$ is the set of all possible time windows. $duration \in \mathbb{U}_{tw} \to \mathbb{R}$ maps a time window to a real valued representation of the difference between the its start and end in the granularity of seconds.*

For $tw = (t_s, t_e)$, $\pi_s(tw) = t_s$ and $\pi_e(tw) = t_e$. For instance, $tw_1 =$ (2022-01-01 00:00:00, 2022-01-08 00:00:00) is a time window where $\pi_s(tw_1) =$ 2022-01-01 00:00:00, $\pi_e(tw_1) =$ 2022-01-08 00:00:00, and $duration(tw_1) = 604800$ (seconds). Note that, in the remainder, we denote 604800 as *week*.

A time span of an event log with length l is a collection of non-overlapping time windows of the event log that have the equal duration of l.

Definition 5 (Time Span). *Let $l \in \mathbb{R}$ be a time span length. Let $L \in \mathbb{U}_e^*$ be an event log. $t_{min}(L) = min_{e \in E(L)} \#_{time}(e)$, $t_{max}(L) = max_{e \in E(L)} \#_{time}(e)$, and $n_l(L) = \lceil (t_{max}(L) - t_{min}(L))/l \rceil$. $span_l(L) = \{(t_{min}(L) + (k-1) \cdot l, t_{min}(L) + k \cdot l) \mid 1 \leq k \leq n_l(L)\}$. For any $e \in E(L)$, $tw_{l,L}(e) = tw$ s.t. $\pi_s(tw) \leq \#_{time}(e) \leq \pi_c(tw)$.*

Assume that event log L contains traces that consist of events between 2022-01-01 00:00:00 and 2022-01-15 00:00:00. $t_{min}(L) =$ 2022-01-01 00:00:00, $t_{max}(L) =$ 2022-01-15 00:00:00, and $n_{week}(L) = 2$. $span_{week}(L)$ contains two time windows $tw_1 =$ (2022-01-01 00:00:00, 2022-01-08 00:00:00) and $tw_2 =$ (2022-01-08 00:00:00, 2022-01-15 00:00:00).

4   Context-Aware Deviation Detection

In this section, we introduce a context-aware deviation detection problem and explain an ontology of contexts for context-aware deviation detection.

4.1 Context-Aware Deviation Detection Problem

First, a deviation detection problem is to compute a function that labels traces either with label *deviating* or with label *normal*. All known deviation detection methods implicitly or explicitly use some form of scoring of traces *score* that is a mapping of traces to some real number (cf. Subsect. 2.1). A threshold τ is used to decide the label. We conceptualize deviating traces as traces scored above τ.

Definition 6 (Deviation Detection). *Let L be an event log. Let $\mathbb{S} = [0,1]$ be a range of all possible score values and $\tau \in \mathbb{S}$ be a threshold value. A score function score $\in L \rightarrow \mathbb{S}$ maps traces to score values. $detect_{score} \in L \rightarrow \{d, n\}$ is a deviation detection using score such that, for any $\sigma \in L$, $detect_{score}(\sigma) = d$ if $score(\sigma) > \tau$. $detect_{score}(\sigma) = n$ otherwise.*

Instead of the two-class labeling problem, a context-aware deviation detection problem is a four-class labeling problem. Table 1 describes the four classes with two dimensions: *non-context* and *context*. The non-context deviating (d) and normal (n) correspond to the two classes of the deviation detection problem, whereas context-deviating (d_c) and context-normal (n_c) indicate that a trace is deviating and normal, respectively, when considering context. First, *context-insensitive deviating* (i.e., $d \rightarrow d_c$) indicates that a trace is both non-context deviating and context-deviating. Second, *context-sensitive deviating* (i.e., $n \rightarrow d_c$) denotes that a trace is non-context normal, but context-deviating. Third, *context-sensitive normal* (i.e., $d \rightarrow n_c$) indicates that a trace is non-context deviating, but context-normal. Finally, *context-insensitive normal* (i.e., $n \rightarrow n_c$) denotes that a trace is both non-context normal and context-normal.

Table 1. Four classes in a context-aware deviation detection problem

$\sigma \in \mathbb{U}_e^*$		Context	
		Deviating (d_c)	Normal (n_c)
Non-context	Deviating (d)	Context-insensitive deviating $(d \rightarrow d_c)$	Context-sensitive normal $(d \rightarrow n_c)$
	Normal (n)	Context-sensitive deviating $(n \rightarrow d_c)$	Context-insensitive normal $(n \rightarrow n_c)$

Definition 7 (Context-Aware Deviation Detection Problem). *Given $L \subseteq \mathbb{U}_e^*$, compute a function that labels traces with context-insensitive deviating, context-sensitive deviating, context-sensitive normal, or context-insensitive normal, i.e., c-detect $\in L \rightarrow \{d \rightarrow d_c, n \rightarrow d_c, d \rightarrow n_c, n \rightarrow n_c\}$.*

4.2 Context-Awareness

Based on existing work on contexts of business processes introduced in Subsect. 2.2, we provide context ontology for context-aware deviation detection in Fig. 2. First, *intrinsic* context is inherent to an event. The intrinsic contexts *resource* and *data* correspond to the organizational and data perspectives for a single event. The *waiting time* context represents the average waiting time of an event.

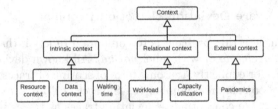

Fig. 2. An ontology of business process context for deviation detection [2,7,9,16].

Thus, the information of *waiting time* contexts can be used to capture unusually long delays for events.

Next, *relational* context is inherent to the relation of multiple events. The relational contexts *workload, waiting time* and *capacity utilization* represent context information that is measured (extracted) by relating multiple events of the data. The *workload* context represents event counts of various selections for a given time window. The *capacity utilization* context represents workloads of resources or locations of events by counting the respective events that were recorded during the time window of the context, e.g., the capacity utilization of a finance department. Therefore, the information of *capacity utilization* contexts can be used to capture unusually high workloads of resources.

Finally, *external* context is not directly attributable to events, but still affects them. The external context *pandemics* represents the outbreak of infectious disease, e.g., COVID-19 pandemic. As an external context is not directly measurable on event data, either additional data has to be used, or it has to be represented by another measurable relational context caused by the external context, e.g., a hygienic products shop experiences exceptionally large demand during the first worldwide outbreak of Corona pandemic such that the *workload* context captures the unusual demand increase and, thus, the external context *pandemic*.

5   Framework for Context-Aware Deviation Detection

This section introduces a framework based on *post-processing mechanism*. We explain each of the four components described in Fig. 1 with a running example: 1) deviation detection, 2) context analysis, 3) context link, and 4) post processing.

5.1   Running Example

Figure 3 shows a running example of an order management process. It describes events of the process for two weeks under 1) the context of high workload (i.e., many events during the week) in *week 1* and 2) the context of overwork (i.e., many events during the weekend) in *week 2*. The context of high workload is considered as a positive context, i.e., the context justifies deviating traces in *week 1*, producing more context-normal traces. In contrast, we consider the context of overwork as a negative context, i.e., the context refutes normal traces in *week 2*, producing more context-deviating traces in *week 2*.

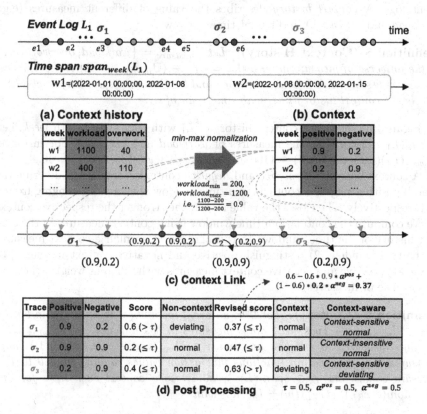

Fig. 3. A running example of context-aware deviation detection for the time window *week 1 (w1)* and *week 2 (w2)*. **(a)** The context history of L_1 in *w1* shows *workload* of 1100 (total number of events in *w1*) and *overwork* of 40 (total number of events during weekend in *w1*), respectively. **(b)** Assume *workload* is a positive measure, $workload_{max} = 1200$, and $workload_{min} = 200$. By aggregating positive (blue) and negative (red) measures in *w1* with *min-max normalization*, we compute the context in *w1*, i.e., positive context of 0.9 and negative context of 0.2. **(c)** We first connect the context to events (as denoted by gray dotted lines) and then connect the context to a trace by computing the maximum positive and negative contexts of its events. σ_2 has the positive context of 0.9 (i.e., the maximum positive context of its events) and the negative context of 0.9 (i.e., the maximum negative context). **(d)** The non-context deviating score of σ_1 is 0.6 ($> \tau$, i.e., non-context deviating), but its revised deviation score is 0.37 ($\leq \tau$, i.e., context-normal). Thus, σ_1 is context-sensitive normal.

5.2   Context Analysis

We analyze context in two steps. First, we compute *context history* based on event logs. A *context history* describes the value of different measures (e.g., *workload* and *overwork*) in different time windows.

Definition 8 (Context History). *Let* $\mathbb{U}_{measure} = \{workload, overwork, \dots\}$ *be the universe of measure names.* $\mathbb{U}_{ch} = \mathbb{U}_{tw} \nrightarrow (\mathbb{U}_{measure} \nrightarrow \mathbb{R})$ *is the universe of context history. Let* L *be an event log and* $l \in \mathbb{R}$ *a time span length.* $ch_l(L) \in \mathbb{U}_{ch}$ *is the context history in* L *with time span of* l.

Figure 3(a) shows the context history of L_1 with time span length *week*, i.e., $ch_{week}(L_1)$. It contains the measures of *workload* and *overwork*. For instance, $ch_{week}(L_1)(w1)(workload) = 1100$ and $ch_{week}(L_1)(w1)(overwork) = 40$.

A context consists of *positive* and *negative* context scores. They describe the overall positive/negative contexts in a time window with a value ranging from 0 to 1, respectively. The closer the value is to 1, the stronger the respective context is. We compute the context in a time window using context measures in the context history of the time window. To this end, we 1) normalize context measures in the time window, 2) distinguish positive and negative context measures, 3) aggregate positive and negative context measures with different weights (i.e., the importance of measures).

Definition 9 (Context). *Let* L *be an event log and* $l \in \mathbb{R}$ *a time span length.* $type \in \mathbb{U}_{measure} \rightarrow \{pos, neg\}$ *maps measures to pos and neg,* $w \in \mathbb{U}_{measure} \rightarrow \mathbb{R}$ *maps measures to weights, and* $norm \in \mathbb{U}_{measure} \rightarrow (\mathbb{R} \rightarrow [0,1])$ *maps measures to normalization functions that assign values ranging from 0 to 1 to measure values.* $ctx_{l,L} \in span_l(L) \nrightarrow [0,1]^2$ *is a context such that, for any* $tw \in dom(ctx_{l,L})$, $ctx_{l,L}(tw) = (pc, nc)$ *with*

- $pc = \sum_{m \in dom(ch_{l,L}^{tw}) \wedge type(m)=pos} w(m) \cdot norm(m)(ch_{l,L}^{tw}(m))/w(m)$ *and*
- $nc = \sum_{m \in dom(ch_{l,L}^{tw}) \wedge type(m)=neg} w(m) \cdot norm(m)(ch_{l,L}^{tw}(m))/w(m)$

, where $ch_{l,L}^{tw} = ch_l(L)(tw)$.

The example in Fig. 3 assumes $norm_1$, $type_1$, and w_1. First, $norm_1$ uses *min-max normalization* for each measure, e.g., with the maximum *workload* of 1200, the minimum *workload* of 200, the maximum *overwork* of 120, and the minimum *overwork* of 20. Moreover, $type_1$ classifies *workload* as a positive context measure and *overwork* as a negative context measure, i.e., $type_1(workload) = pos$ and $type_1(overwork) = neg$. Finally, w_1 assigns the weights of 10 and 5 to *workload* and *overwork*, respectively, i.e., $w_1(workload) = 10$ and $w_1(overwork) = 5$.

Figure 3(b) shows context ctx_{week,L_1}. The positive context in time window $w1$ is $w_1(workload) \cdot norm_1(workload)(1200)/w_1(workload) = 10 \cdot 0.9/10 = 0.9$. The negative context in $w1$ is $w_1(overwork) \cdot norm_1(overwork)(200)/w_1(overwork) = 5 \cdot 0.2/5 = 0.2$. Note that, in the example, the weight does not play its role since we only use one positive and one negative context measure.

5.3   Linking Context to Traces

To connect context to traces, we first link context to events. An event is connected to the context of the time window that the event belongs to.

Definition 10 (Context-Event Link). *Let L be an event log and $l \in \mathbb{R}$ a time span length. A context-event link, $elink_{l,L} \in E(L) \to [0,1]^2$, maps events to positive and negative contexts such that, for any $e \in E(L)$, $elink_{l,L}(e) = ctx_{l,L}(tw_{l,L}(e))$.*

As depicted in Fig. 3(c) by gray dotted lines, $e1$, $e2$, and $e3$ by σ_1 and $e4$ and $e5$ by σ_2 are connected to $ctx_{week,L_1}(w1)$, i.e., $elink_{week,L_1}(e1) = ctx_{week,L_1}(w1) = (0.9, 0.2)$, etc.

The context of a trace is determined by the context of its events. In this work, we define the maximum positive and negative context of the events of a trace as the context of the trace.

Definition 11 (Context-Trace Link). *Let L be an event log and $l \in \mathbb{R}$ a time span length. $tlink_{l,L} \in L \to [0,1]^2$ maps traces to positive and negative contexts s.t., for any $\sigma \in L$, $tlink_{l,L}(\sigma) = (\max(\{pc \in [0,1] \mid \exists_{e \in elem(\sigma)} (pc,nc) = elink_{l,L}(e)\}), \max(\{nc \in [0,1] \mid \exists_{e \in elem(\sigma)} (pc,nc) = elink_{l,L}(e)\}))$.*

As shown in Fig. 3(c), σ_1 has the positive context of 0.9 and negative context of 0.2, i.e., $tlink_{week,L_1}(\sigma_1) = (0.9, 0.2)$, since the maximum positive context of its events, i.e., e_1, e_2, and e_3, is 0.9 and the maximum negative context is 0.2. $tlink_{week,L_1}(\sigma_2) = (0.9, 0.9)$, since the maximum positive context of its events, i.e., e_4, e_5, and e_6, is 0.9 and the maximum negative context is 0.9.

5.4   Post Processing

Post-processing function revises the non-context deviating score of a trace using the positive and negative context of the trace. The positive context decreases the deviating score, whereas the negative context increases it.

Definition 12 (Post Processing). *Let L be an event log, $l \in \mathbb{R}$ a time span length, and score a score function. $post_{l,L,score} \in L \times [0,1]^2 \to [0,1]$ maps a trace, a positive degree, and a negative degree to revised score such that, for any $\sigma \in L$, $\alpha^{pos} \in [0,1]$, and $\alpha^{neg} \in [0,1]$, $post_{l,L,score}(\sigma, \alpha^{pos}, \alpha^{neg}) = score(\sigma) - score(\sigma) \cdot \alpha^{pos} \cdot pc + (1 - score(\sigma)) \cdot \alpha^{neg} \cdot nc$ where $(pc, nc) = tlink_{l,L}(\sigma)$.*

In Fig. 3(d), σ_1 has the deviation score of 0.6, i.e., $score_1(\sigma_1) = 0.6$. Given σ_1, $\alpha^{pos} = 0.5$ and $\alpha^{neg} = 0.5$, $post_{week,L_1,score_1}$ revises the deviating score to a new score of 0.37, i.e., $0.6 - 0.6 \cdot 0.5 \cdot 0.9 + (1 - 0.6) \cdot 0.5 \cdot 0.2 = 0.37$.

Finally, a context-aware detection function labels traces with the four context-aware classes described in Table 1, based on the non-context deviating score and revised deviating score.

Definition 13 (Context-Aware Detection). *Let L be an event log and $l \in \mathbb{R}$ a time span length. Let score be a score function. Let $\alpha^{pos}, \alpha^{neg} \in [0,1]$ be positive and negative degrees and $\tau \in \mathbb{S}$ be a threshold. c-detect $\in L \to \{d \to d_c, n \to d_c, d \to n_c, n \to n_c\}$ maps traces to context-aware labels such that for any $\sigma \in L$:*

$$c\text{-}detect(\sigma) = \begin{cases} d \to d_c & \text{if } detect_{score}(\sigma) = d \text{ and } post_{l,L,score}(\sigma, \alpha^{pos}, \alpha^{neg}) > \tau \\ n \to d_c & \text{if } detect_{score}(\sigma) = n \text{ and } post_{l,L,score}(\sigma, \alpha^{pos}, \alpha^{neg}) > \tau \\ d \to n_c & \text{if } detect_{score}(\sigma) = d \text{ and } post_{l,L,score}(\sigma, \alpha^{pos}, \alpha^{neg}) \le \tau \\ n \to n_c & \text{if } detect_{score}(\sigma) = n \text{ and } post_{l,L,score}(\sigma, \alpha^{pos}, \alpha^{neg}) \le \tau \end{cases}$$

As shown in Fig. 3(d), given $\tau = 0.5$, $\alpha^{pos} = 0.5$, and $\alpha^{neg} = 0.5$, c-detect $(\sigma_1) = d \to n_c$ since $detect_{score_1}(\sigma_1) = d$ and $post_{week,L_1,score_1}(\sigma_1, \alpha^{pos}, \alpha^{neg}) = 0.37 \le \tau$. Furthermore, c-detect$(\sigma_3) = n \to d_c$ since $detect_{score_1}(\sigma_2) = n$ and $post_{week,L_1,score_1}(\sigma_2, \alpha^{pos}, \alpha^{neg}) = 0.63 > \tau$.

6   Implementation

The framework for context-aware deviation detection is implemented as a cloud-based web service with a dedicated user interface. The implementation is available at https://github.com/janikbenzin/contect along with the source code, a user manual, and a demo video. It consists of four functional components: (1) context analysis, (2) deviation detection, (3) context-aware deviation detection, and (4) visualization.

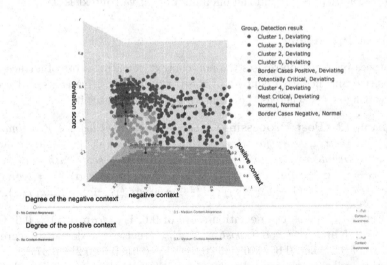

Fig. 4. A screenshot of *Scatter* visualization. By varying the degree of positive and negative context, we can deduce the adequate degree of positive and negative context to be used for the context-aware deviation detection.

First, the context analysis component supports the computation of the context history and context. The context introduced in Fig. 2 have been implemented including *workload, weekend, waiting time,* and *capacity utilization.*

Second, the deviation detection component implements four deviation detection methods that correspond to representatives of four respective categories introduced in Subsect. 2.1. For process-centric methods, we adapt the two-step approach in [8] by using *Inductive* miner [10] for process discovery and *alignment* [1] for conformance checking. For profile-based approaches, *Profiles* [11] has been implemented, while *ADAR* [6] and *Autoencoder* [13] have been implemented as process-agnostic & interpretable/non-interpretable approaches, respectively. Next, the context-aware deviation detection component implements the post processing and the context-aware deviation detection function.

Finally, the visualization component supports an analysis view for each deviation detection method. Each analysis view consists of three visualizations: *tabular, scatter,* and *calendar. Tabular* visualizes the most deviating traces by sorting them based on the deviation score, the proximity to being relabelled as context-normal, etc. *Scatter* shows a 3D-scatter plot of the deviation score, positive context, and negative context, as shown in Fig. 4. As the number of deviating traces can be large, the k-Medoids clustering algorithm is applied to all deviating traces such that the user can analyze the medoids to understand the whole space of deviating traces more efficiently (depicted as first to fourth and seventh legend entry in Fig. 4). Moreover, by varying the positive and negative degrees, we can analyze the effect of the context on the deviation detection. *Calendar* visualizes the context over time by aggregating contexts by time and plotting them over the time span.

7 Evaluation

This section evaluates the proposed framework using the implementation in Sect. 6. To this end, we conduct four case studies using deviation detection methods: *Inductive, Profiles, ADAR,* and *Autoencoder.* In each case study, we compare the performance of context-aware deviation detection and context non-aware deviation detection in 225 different simulated scenarios. In the rest of this section, we first introduce a detailed experimental design and then report the results.

7.1 Experimental Design

As depicted in Fig. 5, the evaluation follows a four step pipeline: *data generation, simulation scenario injection, framework application,* and *evaluation of results.*

First, the data generation uses CPN Tools[1] to simulate an order management process. Next, we inject four different types of deviating events into the generated event data and label them as non-context deviating: 1) *Rework* randomly adds an event to a trace with the activity that has already occurred, 2) *Swap*

---

[1] www.cpntools.org.

randomly swaps the timestamp of two existing events, 3) *Replace resource* randomly replaces the resource of an event with a different resource, and 4) *Remove* randomly removes an existing event from the data. To understand the effect of the amount of deviations on the classification result, the evaluation injected 2%, 5%, or 10% deviations equally distributed among the four types.

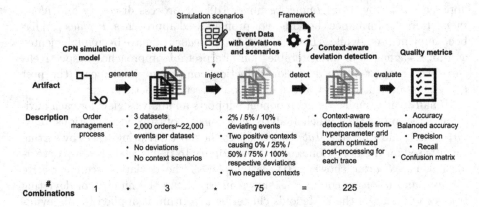

Fig. 5. An overview of the experimental design

Afterward, we inject four contextual scenarios as follows.

1. For *workload* scenario, we randomly select a week and add additional orders in the week. We consider it as a positive context and, thus, the non-context deviating events of the selected week are relabelled to context-normal.
2. For *capacity utilization performance* scenario, we randomly assign vacations and sick leaves to resources, lowering the capacity of the process. It is considered as a positive context, and non-context deviating events associated with the reduced capacity resource are relabelled to context-normal.
3. For *waiting time*, all events of randomly chosen days are randomly delayed. It is considered a negative context, and all of the delayed events that are non-context normal or context-normal are labeled as context-deviating.
4. For *overwork* scenario, we shift the random percentage of events during weekdays to Saturday and Sunday. It is regarded as a negative context, and all shifted events that are non-context normal or context-normal are relabelled to context-deviating.

To determine the strength of the relationship between positive contexts and deviations, we use *% context attributable* parameter that determines how many traces are affected by positive contextual scenarios, i.e., non-context deviating events are relabelled to context-normal. We include it as the second parameter for experiments with values ranging from 0% to 100% as depicted in Fig. 5.

225 experiments per case study ($3 * 3 * 5 * 5$) result from the parameters as shown in Fig. 5, i.e., three event datasets, three % events deviating parameters and the five % context attributable parameters per positive contextual scenario.

Next, we apply the proposed framework and compute context-aware detection results. Hyperparameter grid search is applied to find the best combination of positive and negative degrees for the post function.

Table 2. Evaluation results from four case studies

		Context-non-aware deviation detection $\alpha^{pos} = \alpha^{neg} = 0$	Context-aware deviation detection $\alpha^{pos}, \alpha^{neg}$ optimized	Difference
Inductive	Accuracy	0.389118	0.426846	−0.037728
	Avg. class accuracy	0.326856	0.311832	−0.015024
	Precision	0.248691	0.293496	−0.044805
	Recall	0.389118	0.426846	−0.037728
Autoencoder	Accuracy	0.385035	0.425686	−0.040651
	Avg. class accuracy	0.311249	0.312451	−0.001202
	Precision	0.235668	0.369101	−0.133433
	Recall	0.385035	0.424996	−0.039961
Profiles	Accuracy	0.363995	0.406368	−0.042373
	Avg. class accuracy	0.293880	0.292083	−0.001797
	Precision	0.220972	0.332658	−0.111686
	Recall	0.363995	0.404011	−0.034061
ADAR	Accuracy	0.351544	0.395066	−0.043522
	Avg. class accuracy	0.291969	0.289284	−0.002685
	Precision	0.229760	0.334021	−0.104261
	Recall	0.351544	0.385152	−0.033608

7.2 Experimental Results

First, we report average results for each case study in Table 2, showing that the consideration of positive/negative context is effective in the context-aware deviation detection. The first column in Table 2 shows the performance of context-non-aware deviation detection with α^{pos} and α^{neg} both set to 0. The second column in Table 2 shows the performance of context-aware deviation detection with positive α^{pos} and negative degree α^{neg} both optimized through the hyperparameter grid search. The third column shows the performance difference of the proposed approach with respect to the baseline.

In the case study using *Inductive*, the accuracy of *0.389118* is improved by *0.037728* to *0.426846*, the average class accuracy of *0.326856* is slightly reduced by *0.015024* to *0.311832*, the precision of *0.248691* is boosted by *0.044805* to *0.293496* and the recall of *0.389118* is upgraded by *0.037728* to *0.425686*. The other three case studies also show performance improvements in terms of accuracy, precision, and recall similar to *Inductive* and a decrease in average class accuracy. In particular, the results are significantly more precise with the framework's context-aware deviation detection than for deviation detection.

Second, Fig. 6 shows two confusion matrices in Fig. 6 for *Inductive* and *Autoencoder*, summing the confusion matrix of each experiment. The confusion matrix for *Autoencoder* is representative for *Profiles* and *ADAR*, showing similar results. The context-awareness generally improves the performance in all case studies by improving the detection of *context-sensitive deviating* traces, but not by detection of *context-sensitive normal* traces. With respect to *context-sensitive normal*, the framework's context-awareness has most of the time does not correctly predict the *context-sensitive normal* traces (0 out of *9,194 + 9,389 + 2,798 = 21,381 context-sensitive normal* traces for *Inductive* and *83* out of

$6,306 + 11,173 + 83 + 3,678 = 21,240$ traces for *Autoencoder*). With respect to *context-sensitive deviating*, the framework's context-awareness performs significantly better for the *context-sensitive deviating* traces with $54,186$ of $72,021 + 36,952 + 0 + 54,187 = 163,160$ correctly predicted traces (*Inductive*) and with $47,951$ of $50,485 + 60,529 + 452 + 47,951 = 159,417$ correctly predicted traces (*Autoencoder*).

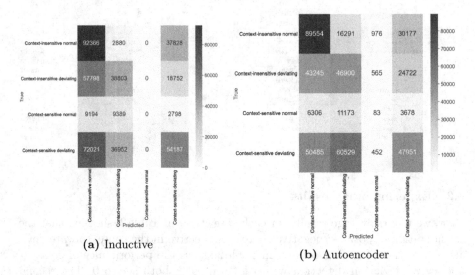

(a) Inductive   (b) Autoencoder

Fig. 6. Confusion matrices summed over all 225 experiments of the respective context-aware deviation detection method

8   Conclusion

In this paper, we proposed a framework to support context-aware deviation detection. The proposed framework can incorporate any existing unsupervised deviation detection methods with varying strengths and weaknesses and enhance them with various contextual aspects. We have implemented the framework as an extensible web service with a dedicated user interface. Moreover, we have evaluated the effectiveness of the framework by conducting experiments using representative deviation detection methods in different contextual scenarios.

This work has several limitations. First, the proposed framework introduces several parameters that possibly affect the detection results, e.g., the negative and positive degree of *post* function, the threshold of *score* function, etc. Second, the framework is dependent on the performance of the deviation detection method. Third, using an event log as the input, the framework only indirectly measures external contexts.

Besides addressing the above limitations, in future work, we plan to extend the framework to support the root cause analysis of context-aware deviations. We can analyze the relevant context of context-aware deviating instances and trace back the relevant context measure, e.g., high workload. Moreover, we plan to extend the framework to consider contexts of different time window lengths, e.g., context in *week*, *day*, and *hour*. Another direction of future work is to develop different post functions to improve the performance of the context-aware deviations.

References

1. van der Aalst, W.M.P., Adriansyah, A., van Dongen, B.F.: Replaying history on process models for conformance checking and performance analysis. Wiley Interdiscip. Rev. Data Min. Knowl. Discov. **2**(2), 182–192 (2012)
2. van der Aalst, W.M.P., Dustdar, S.: Process mining put into context. IEEE Internet Comput. **16**(1), 82–86 (2012)
3. Bezerra, F., Wainer, J.: Algorithms for anomaly detection of traces in logs of process aware information systems. Inf. Syst. **38**(1), 33–44 (2013)
4. Böhmer, K., Rinderle-Ma, S.: Anomaly detection in business process runtime behavior - challenges and limitations. CoRR abs/1705.06659 (2017)
5. Böhmer, K., Rinderle-Ma, S.: Multi instance anomaly detection in business process executions. In: Carmona, J., Engels, G., Kumar, A. (eds.) BPM 2017. LNCS, vol. 10445, pp. 77–93. Springer, Cham (2017). https://doi.org/10.1007/978-3-319-65000-5_5
6. Böhmer, K., Rinderle-Ma, S.: Mining association rules for anomaly detection in dynamic process runtime behavior and explaining the root cause to users. Inf. Syst. **90**, 101438 (2020)
7. Dockhorn Costa, P., Almeida, J.P.A., Ferreira Pires, L., van Sinderen, M.: Situation specification and realization in rule-based context-aware applications. In: Indulska, J., Raymond, K. (eds.) Distributed Applications and Interoperable Systems, pp. 32–47 (2007)
8. Jalali, H., Baraani, A.: Genetic-based anomaly detection in logs of process aware systems. World Acad. Sci. Eng. Technol. **64**(4), 304–309 (2010)
9. Kronsbein, D., Meiser, D., Leyer, M.: Conceptualisation of contextual factors for business process performance. Lecture Notes in Engineering and Computer Science 2210 (2014)
10. Leemans, S.J.J., Fahland, D., van der Aalst, W.M.P.: Discovering block-structured process models from event logs containing infrequent behaviour. In: Lohmann, N., Song, M., Wohed, P. (eds.) BPM 2013. LNBIP, vol. 171, pp. 66–78. Springer, Cham (2014). https://doi.org/10.1007/978-3-319-06257-0_6
11. Li, G., van der Aalst, W.M.P.: A framework for detecting deviations in complex event logs. Intell. Data Anal. **21**(4), 759–779 (2017)
12. Mannhardt, F., de Leoni, M., Reijers, H.A., van der Aalst, W.M.P.: Balanced multi-perspective checking of process conformance. Computing **98**(4), 407–437 (2015). https://doi.org/10.1007/s00607-015-0441-1
13. Nolle, T., Luettgen, S., Seeliger, A., Mühlhäuser, M.: Analyzing business process anomalies using autoencoders. Mach. Learn. **107**(11), 1875–1893 (2018). https://doi.org/10.1007/s10994-018-5702-8
14. Nolle, T., Luettgen, S., Seeliger, A., Mühlhäuser, M.: BINet: Multi-perspective business process anomaly classification. CoRR abs/1902.03155 (2019)
15. Pauwels, S.: An anomaly detection technique for business processes based on extended dynamic Bayesian networks. In: Proceedings of the ACM Symposium on Applied Computing Part, F1477, pp. 494–501 (2019)

16. Song, R., Vanthienen, J., Cui, W., Wang, Y., Huang, L.: Towards a comprehensive understanding of the context concepts in context-aware business processes. In: Betz, S. (ed.) S-BPM ONE 2019, pp. 5:1–5:10 (2019)
17. Warrender, C., Forrest, S., Pearlmutter, B.A.: Detecting intrusions using system calls: Alternative data models. In: 1999 IEEE Symposium on Security and Privacy, pp. 133–145 (1999)
18. Ye, J., Dobson, S., McKeever, S.: Situation identification techniques in pervasive computing: a review. Pervasive Mob. Comput. 8(1), 36–66 (2012)

When to Intervene? Prescriptive Process Monitoring Under Uncertainty and Resource Constraints

Mahmoud Shoush[ID] and Marlon Dumas[✉][ID]

University of Tartu, Tartu, Estonia
{mahmoud.shoush,marlon.dumas}@ut.ee

Abstract. Prescriptive process monitoring approaches leverage historical data to prescribe runtime interventions that will likely prevent negative case outcomes or improve a process's performance. A centerpiece of a prescriptive process monitoring method is its intervention policy: a decision function determining if and when to trigger an intervention on an ongoing case. Previous proposals in this field rely on intervention policies that consider only the current state of a given case. These approaches do not consider the tradeoff between triggering an intervention in the current state, given the level of uncertainty of the underlying predictive models, versus delaying the intervention to a later state. Moreover, they assume that a resource is always available to perform an intervention (infinite capacity). This paper addresses these gaps by introducing a prescriptive process monitoring method that filters and ranks ongoing cases based on prediction scores, prediction uncertainty, and causal effect of the intervention, and triggers interventions to maximize a gain function, considering the available resources. The proposal is evaluated using a real-life event log. The results show that the proposed method outperforms existing baselines regarding total gain.

Keywords: Prescriptive process monitoring · Causal inference · Uncertainty

1 Introduction

Prescriptive Process monitoring (PrPM) is a family of process mining methods that trigger runtime actions to optimize a process's performance [6,14]. PrPM methods use *event logs* describing past business process executions to train *machine learning (ML)* algorithms for two goals. *First,* the trained ML models predict how an instance of the process (a.k.a. case) will unfold. For example, whether the case leads to a positive outcome (e.g., a customer is satisfied) or a negative outcome (e.g., a customer launches a complaint) [20]. *Second,* PrPM methods use ML to assess the effect of triggering an action (herein called an *intervention*) on the probability of a negative outcome or a performance measure.

Supported by the European Research Council (PIX Project).

Recently, various PrPM methods have been proposed [2,6,14,18]. These methods, however, implement intervention policies based on predictions of negative outcomes without considering the uncertainty of these predictions. Also, they trigger an intervention when the predicted probability of a negative outcome is above a threshold, without considering potential increases or decreases in this predicted probability that may occur as the case unfolds further. Finally, these methods do not consider the fact that there are limited resources available to perform the recommended interventions.

In this paper, we address the following problem: Given a set of cases, and given a type of intervention that generally decreases the likelihood of a negative case outcome. *How do we select the cases for which applying the intervention (now or later) maximizes a gain function, considering the available resources to perform interventions?* Here, the gain function considers the tradeoff between the cost of applying the intervention to a case and the cost of negative outcomes.

To address this problem, we first apply an ensemble-based predictive model to estimate the negative outcome probability for each case, and we estimate the associated uncertainty. Using a causal model, we then determine the causal effect of applying an intervention on the negative outcome probability. We then use the negative outcome probability, the uncertainty, and the estimated causal effect and apply a *filtering* and *ranking* mechanism to identify cases for which an intervention would be most profitable (highest gain). We also consider the tradeoff (i.e., opportunity cost) between triggering an intervention in the current state, given the level of uncertainty of the predictive model, versus postponing the intervention to a later state. The paper reports an empirical evaluation comparing the proposed approach against state-of-the-art baselines.

The following section motivates the proposed method. Section 3 then presents background concepts and related work. Section 4 explains the proposed method, while Sect. 5 discusses the empirical evaluation. Finally, Sect. 6 concludes and discusses future work directions.

2 Motivating Example

In *a loan origination process*, a case starts when a customer submits his documents to obtain a loan. Then a process worker (or an employee) verifies the submitted documents. When they are valid, the employee sends an offer to the customer via different channels, such as phone calls or emails. This case ends positively when the customer accepts the offer and receives the loan, or negatively when the customer declines the offer or the employee rejects the application.

The principal concern arises when cases end negatively, leading to less payoff. One way to deal with this could be to predict negative cases based on prediction scores. Then trigger an alarm to take a proactive action or an intervention, e.g., *making a follow-up call*, when the prediction score exceeds a certain threshold.

However, this strategy could be ineffective. Suppose an intervention policy where interventions are triggered to cases that are likely to end negatively based on low-quality prediction scores. Also, all employees are occupied and cannot immediately perform the intervention for all cases. Additionally, triggering interventions without considering their effect could be misleading since they may provide low or negative impact when utilized.

A more proper method is to quantify the prediction uncertainty to estimate how sure predictive models are with the prediction scores. Moreover, measuring the causal effect of utilizing interventions and considering the availability of resources. Another step that may enhance the overall payoff could be considering the tradeoff between triggering interventions now versus postponing them for a later state. In this paper, we discuss this method and evaluate its performance.

3   Background and Related Work

3.1   Predictive Process Monitoring

Predictive process monitoring (PPM) [12] is a complementary set of process mining methods to predict how ongoing cases will end. A PPM technique may, for instance, predict the remaining time for an ongoing case to be executed entirely [22], the following action or activity to be executed [15], or the outcome w.r.t group of outcomes, e.g., positive or negative [20]. This paper focuses on the latter technique, known as an *outcome-oriented* PPM.

Recent outcome-oriented PPM methods estimate the prediction scores, i.e., probability of negative outcomes, for ongoing cases and classify them positively or negatively. If the prediction scores exceed a threshold, e.g., above 0.5, the ongoing case is considered more likely to end negatively.

However, outcome-oriented PPM methods focus only on making predictions as accurate as possible, regardless of the quality of the predictions. These methods rely on several *case bucketing techniques* [4], e.g., a single bucket where cases are made in the same bucket and train one ML algorithm instead of several. Also, they rely on various *feature encoding* techniques [20] to map each case into a feature vector to train the ML algorithm. For instance, an aggregate encoding in which all events from the beginning of the case are considered. Thus, several aggregate functions may be used to the values an event has carried throughout the case. Also, a handful of possible *inter-case* features are extracted [10] to enrich the training of ML or deep learning (DL) algorithms [11]. Still, these techniques aim to improve the performance of the prediction scores and ignore quantifying the prediction quality via measuring *prediction uncertainty*.

To the best of our knowledge, only one work from the literature considers estimating the model's prediction uncertainty explicitly, tackling another PPM task, i.e., remaining time [24]. They learn the prediction uncertainty with artificial neural networks and a Monte Carlo (MC) [7] dropout technique that is unreliable for *out-of-distribution* data (i.e., where there is an input case from a region very far from the trained data) and computationally expensive [1].

Metzger et al. [14] introduces an approach to measure the reliability of prediction scores using an ensemble of DL classifiers at different process states. This approach does not discuss the estimation of the prediction uncertainty ignoring the situation of out-of-distribution input. Moreover, where predictive models provide several prediction scores for the same input, i.e., *outcomes overlap*.

3.2   Prescriptive Process Monitoring

Prescriptive process monitoring (PrPM) methods go beyond predictions to pre-scribe runtime interventions to prevent or mitigate negative outcome effects. These methods aim to improve the performance of business process executions by determining if and when to trigger an intervention to maximize a payoff.

Diverse PrPM methods have been proposed. Metzger et al. [14] suggest using ongoing cases prediction scores and their reliability estimate with a reinforcement learning technique to discover when to trigger runtime interventions. Another work by Fahrenkrog et al. [6] proposes triggering one or more alarms when cases are more likely to end negatively, followed by an intervention.

Both the work of Metzger et al. and that of Fahrenkrog et al. identify cases that need intervention. They assume that resources are unbounded and consider only the current state of a given case to determine when to intervene. Instead, we study the tradeoff between intervening now or later based on the current and future prediction scores. Thus, we identify the most profitable case and assign resources to it, considering that resources are limited.

Weinzerl et al. [23] suggest a PrPM method to recommend the next best activity from a list of possible activities with a higher preference for a pre-defined KPI. Khan et al. [11] introduce a memory-augmented neural net approach to recommend the most suitable path (meaning a set of activities until the comple-tion of the process) based on pre-specified KPIs. Both the work of Weinzerl et al. and that of Khan et al. do not discuss an exact idea of interventions or when to trigger them to maximize payoff.

3.3   Causal Inference

Causal Inference (CI) [25] is a set of methods to predict what would occur if we adjust the process during its execution time by finding a cause-effect relationship between two variables, i.e., an intervention (T) and an outcome (Y).

CI methods mainly unfold into two categories [8]. The first category is *struc-tural causal models (SCMs)*, a multivariate statistical analysis method explor-ing structural relationships between dependent and independent variables. It depends mainly on discovering and building a causal graph by domain experts.

The second category is *potential outcome frameworks* (a.k.a., the Neyman-Rubin Causal Model). A statistical analysis method that does not require a pre-built causal graph like the SCMs and relies on the concept of potential outcomes. We use this category in this paper to automatically estimate the causal effect (or *conditional average treatment effect (CATE)*) of intervention on negative cases instead of manually building causal graphs.

For example, in a loan-origination process, a customer would have a loan if he received an *intervention (T)*, e.g., a follow-up call three days after receiving the first offer; otherwise, he would have a different *outcome (Y)*, e.g., offer declined. Accordingly, to measure the $CATE$ of having a follow-up call, we need to compare Y for the same customer when receiving the follow-up call, i.e., $T = 1$, and not receiving the follow-up call, i.e., $T = 0$.

Recent work uses the potential outcome method to estimate the $CATE$ of utilizing interventions. Specifically, in [2], the Authors introduce a PrPM technique to measure the effect of intervention at an individual case level to reduce the cycle time of the process. It targets another PPM problem and considers only one process execution state; it also assumes that resources are unbounded.

In our previous work [18], we propose a PrPM method that utilizes the potential outcome method and a resource allocator technique in the outcome-oriented PPM to allocate resources to cases with max gain. That work considers only a given case's current state and triggers interventions when the prediction score and the causal effect exceed a threshold. However, we did not discuss the tradeoff between triggering an intervention now, given the level of uncertainty of the underlying predictive models versus later.

4 Approach

The proposed PrPM method consists of two main phases, training and testing, see Fig. 1. In the training phase, we train two ML models, i.e., predictive and causal. While in the testing phase, we present filtering and ranking techniques to determine the most profitable case. Then, decide when to trigger an intervention for the selected case to maximize a gain function-considering ongoing cases' current and future state scores, uncertainty estimation, and resource availability.

4.1 Training Phase

We first prepare the process execution data; then, we construct an ensemble-based predictive model to estimate the prediction scores, i.e., the probability of cases likely to end negatively and quantify the prediction uncertainty. Further, we build a causal model to estimate the $CATE$.

Event Log Preprocessing. This step is vital for PPM or PrPM tasks. In PPM, it includes *data cleaning, prefix extraction, and prefix encoding* see Fig. 2. These steps have been discussed in Teinemaa et al. [20], and we follow their suggestions here. We first pre-process the log to dismiss incomplete cases and then extract length k from every case that results in a so-called *prefix log*. This prefix extraction ensures that our training data is equivalent to the testing data. Finally, we encode each trace prefix into a feature vector (X) to train the predictive ensemble model, see Fig. 2.

While in PrPM, one further step is needed to analyze and understand the data and the business objective to identify an intervention T that could positively impact an outcome Y. Moreover, determining what other variables (W: a.k.a. *confounders*) affect the intervention and outcome.

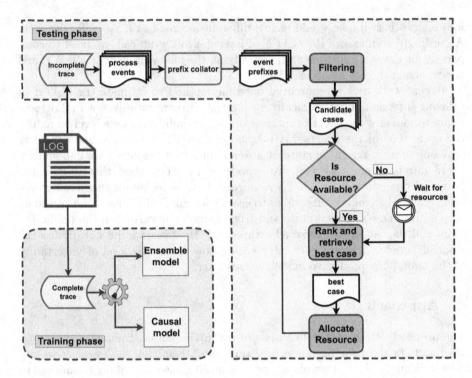

Fig. 1. An overview of the proposed approach.

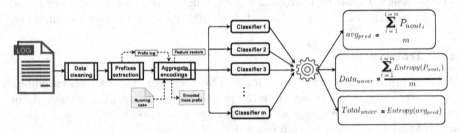

Fig. 2. An overview of the ensemble model.

Ensemble Model. We construct an outcome-oriented predictive model (a classification problem from an ML perspective) via ensemble learning [5], as shown in Fig. 2. The principal assumption of ensemble learning is constructing one robust predictive model from several weak ones. Accordingly, the overall prediction scores performance would be superior where overfitting and the chance of getting a local minimum are avoided, which has two advantages to our work. First, it ensures that we will accurately predict the probability of negative outcomes, i.e., *the prediction score*. Second, it allows estimating *the prediction uncertainty*.

A single classifier in the ensemble is a probabilistic model $cls_i : f(X, P_{uout_i})$. X is the input feature vector, and the P_{uout_i} is the estimated probability of cases likely to end with negative outcomes where $i \in \{1, m\}$, and m is the number of classifiers in the ensemble. We then define the *prediction score* as the average of individual classifiers' prediction scores (avg_{pred}), as shown in Eq. 1. The following step is to estimate the prediction uncertainty.

$$avg_{pred} = \frac{\sum_{i=1}^{m} P_{uout_i}}{m} \tag{1}$$

There are two sources of uncertainties where predictive models become unsure about predictions [9]. The first source is *data (or aleatoric)* uncertainty (σ). It is a property of the distribution that generates cases, and it occurs when *outcomes overlap* or there is *noise* in the underlying data distribution. The second source is *knowledge (or epistemic)* uncertainty (ρ). It is a property of the predictive model's learning parameters and arises due to a lack of model knowledge. It appears when there is *out-of-distribution* input. Both sources form the prediction uncertainty, i.e., *total uncertainty* ($total_{uncer}$), see Eq. 2.

$$total_{uncer} = \sigma + \rho \tag{2}$$

The proper level of σ is defined as *entropy* of the actual underlying data distribution [13]. Here, the entropy is the average level of surprise or uncertainty inherent in the possible outcomes. It is calculated for a random variable S with c in C discrete states, see Eq. 3, where $P(c)$ is the probability of c to occur.

$$Entropy(S) = -\sum_{c \in C} P(c) \log(P(c) \tag{3}$$

However, we do not have access to the actual underlying data distribution, but our model is probabilistic and is trained on this data. So measuring the entropy of our probabilistic model, i.e., trained using *negative log-likelihood,* estimates the level of σ. In particular, we obtain a distribution over outcome labels from individual classifiers in the ensemble for a given case, and σ is the average *entropy* of the individual prediction scores, as shown in Eq. 4.

$$\sigma = \frac{\sum_{i=1}^{m} Entropy(P_{uout_i})}{m} \tag{4}$$

Furthermore, we rely on the *Bayes rule* [19] to estimate the ρ [13]. Assume we obtain a posterior over model parameters that give us the distribution over likely models that have generated the data. Accordingly, models sampled from the obtained distribution agree on data they have seen and provide similar predictions, indicating low ρ. However, if the models do not understand the input, they provide diverse predictions and strongly disagree, indicating high ρ. Thus, ρ is the mutual information between the models' parameters and the predictions.

Similarly, the total uncertainty is the entropy of the average prediction, see Eq. 5. To exemplify the estimation of the prediction uncertainty ($total_{uncer}$), suppose we show the ensemble of classifiers several kinds of input. (1) We give input that all classifiers understand and yield the exact prediction scores. Consequently, classifiers are confident about their prediction scores, and the $total_{uncer}$

is minimum. (2) We show input that all classifiers understand and generate identical predictions but high entropy distribution over outcomes; then, classifiers are uncertain with high σ. (3) We show the ensemble something none of the classifiers understand; hence, all classifiers yield different prediction scores because the input comes from a very far region from the training data. Thus, the ensemble is very diverse with high entropy because we average various probability distributions together, then classifiers are uncertain with high ρ.

$$total_{uncer} = Entropy(avg_{pred}) \tag{5}$$

In addition to the ensemble model that estimates avg_{pred} and $total_{uncer}$ in the training phase, we train a causal model to measure the $CATE$. In particular, we utilize an *orthogonal random forest (ORF)* algorithm because it reasonably deals with high-dimensional variable spaces. Thus, it is beneficial since the process execution logs have numerous event attributes with categorical values.

Estimating the $CATE$ means we evaluate the difference between the probability of a negative outcome if we intervene and if we do not intervene. The higher the differences, the stronger the effect of the intervention. For explanation, we recall the motivating example in Sect. 2, where the goal was to improve the performance of a loan origination process by raising the number of successful applications. Accordingly, we would give customers who are likely to decline the first offer another offer in a way that affects the probability of declining the first offer positively. Then we estimate what would happen for the probability of negative outcome when we send customers a second offer and when we do not.

4.2   Testing Phase

We first use the trained ensemble and causal models to obtain avg_{pred}, $total_{uncer}$, and $CATE$ scores for ongoing cases. Then, we use these scores to *filter* ongoing cases into candidate ones and *rank* them to choose the most profitable case to maximize *a gain* function. We consider that resources are bounded and compare ongoing cases' current and future state scores.

Filtering. At run time, new events of ongoing cases keep coming continuously, and we first collate them via *a prefix collator* to accumulate the sequence of events. Thus, at any point in time, we could have one or multiple ongoing cases that we choose from which one is the most profitable to trigger the intervention. Hence, ongoing cases need to be filtered to minimize the search space.

We use avg_{pred}, $total_{uncer}$, and $CATE$ scores to filter ongoing cases into candidate ones. The essence of these scores varies from one to another. The avg_{pred} gives information about whether ongoing cases are likely to end negatively or not. At the same time, the $total_{uncer}$ shows how sure the model is with its predictions. It ranges from 0 to 1, where the predictive model is entirely certain or uncertain, respectively. Moreover, the $CATE$ is crucial to any PrPM technique, representing the expected impact of utilizing intervention on an ongoing case, e.g., when $CATE$ is above 0, it impacts positively.

All the above mentioned scores (avg_{proba}, $total_{uncer}$, $CATE$) are vital to determining candidate cases from which we will choose the most profitable. However, there are two other critical aspects to determine when to intervene and to define the most profitable case; how the estimated scores will change in the following state and whether resources are available or not.

Future State Scores Estimation. Considering only the current state scores of ongoing cases regardless of investigating what would occur in the future could be misleading. Because if we decide not to intervene, maybe it will be more effective to achieve higher gain when we utilize the intervention later than now, or we will be more sure about the prediction of the outcome and decide not to intervene at all. So, discovering what will happen in the future of ongoing cases allows deciding whether to intervene now or later.

To estimate what will happen in the future state, we predict the avg_{pred}, $CATE$, and $total_{uncer}$ in the future. So for each score, we will have two values: one representing the current state (c_avg_{pred}, c_CATE, c_total_{uncer}) and the other representing the future (f_avg_{pred}, f_CATE, f_total_{uncer}). We follow a technique inspired by the *k-nearest neighbors (KNN)* [16] algorithm to get scores representing the future state, given the degree of *similarity* and *frequency*.

We look at previous cases similar to the ongoing one at the next prefix. For example, in a given case at prefix 4, we want to know what will happen at prefix 5. Then, we capture similar prefixes and define an aggregate score, i.e., the *weighted average*. We consider the degree of *similarity* using Euclidean distance to the current prefix, their *frequency* because higher frequency means more weight and the scores for all similar prefixes. Accordingly, scores representing the future state (f_avg_{pred}, f_CATE, f_total_{uncer}) of ongoing cases are the weighted average from similar previous prefixes. The next step is to use current and future estimates to rank candidate cases and select the most profitable.

Rank and Retrieve the Best Case. To define the most profitable case, we first distinguish between *gain* and *adjusted gain*. The *gain* is the benefits we attain at one state only, either current or future, and it means we estimate two gains, one for the current state (c_gain) and the other for the future state (f_gain). In contrast, the *adjusted gain* (adj_{gain}) is the benefits we attain considering current and future states. We use the adj_{gain} as a decision function to determine whether to intervene now or later.

Triggering interventions may come with benefits and, at the same time, comes at a cost. The costs can vary from one process to another. However, for a given case c_{id}, there is generally a cost for applying the intervention ($cost(c_{id}, T_{i=1})$)) when $T = 1$ and not applying the intervention ($cost(c_{id}, T_{i=0})$) when $T = 0$.

The cost of not applying the intervention describes how much we lose if the negative outcome occurs, and it relies on the avg_{pred} and cost of negative outcomes (c_{uout}), as shown in Eq. 6. In contrast, the cost of utilizing the intervention refers to how much we decrease the probability of negative outcomes considering the intervention cost (c_{T_1}), see Eq. 7, which assumes $CATE$ is reliable. Costs are mainly identified via domain knowledge; however, we assume that the c_{T_1} is less than the c_{uout} to obtain meaningful results.

$$cost(c_{id}, T_{i=0}) = avg_{pred} * c_{uout} \tag{6}$$
$$cost(c_{id}, T_{i=1}) = (avg_{pred} - CATE_1) * c_{uout} + c_{T_1} \tag{7}$$

The corresponding gain $(gain(c_{id}, T_{i=1}))$ from utilizing the intervention on c_{id} is the benefits that allow the highest cost reduction in Eq. 8

$$gain(c_{id}, T_{i=1}) = cost(c_{id}, T_0) - cost(c_{id}, T_{i=1}) \tag{8}$$

We estimate the current state gain $(c\_gain(c_{id}, T_{i=1}))$ using scores from the ensemble and causal models, see Eq. 9. In contrast, the gain for the future state $(f\_gain(c_{id}, T_{i=1}))$ is based on the weighted average scores from previous similar prefixes, see Eq. 10.

$$c\_gain(c_{id}, T_{i=1}) = c\_cost(c_{id}, T_0) - c\_cost(c_{id}, T_{i=1}) \tag{9}$$
$$f\_gain(c_{id}, T_{i=1}) = f\_cost(c_{id}, T_0) - f\_cost(c_{id}, T_{i=1}) \tag{10}$$

Determining the gain for candidate cases' current and future states is vital to define the adjusted gain. To explain the adjusted gain, we first define *an opportunity cost* that measures what we lose when choosing between two or more alternatives-for example, utilizing the intervention now or later. The opportunity cost (opp_{cost}) is the difference between the gain we could achieve in the future state of a given case and the gain in the current state, as shown in Eq. 11.

$$opp_{cost} = f\_gain(c_{id}, T_{i=1}) - c\_gain(c_{id}, T_{i=1}) \tag{11}$$

Given the opportunity cost, we define the *adjusted gain* as the payoff (or gain) we acquire from utilizing the intervention on candidate cases considering current and future states. It is the difference between the $c\_gain(c_{id}, T_{i=1})$ and the opp_{cost}, as shown in Eq. 12. Thus, we define *the most profitable case* as the one with the highest adj_{gain}, which means the lowest opp_{cost}.

$$adj_{gain} = c\_gain(c_{id}, T_{i=1}) - opp_{cost} \tag{12}$$

For example, suppose we filtered ongoing cases into three candidates (see Table 1) eligible for the intervention. Also, there is an available resource to do the intervention; we need to choose which one is more suitable to intervene now or later. If we consider only the gain from the current state, we assign resources to $c_{id} = A$ and treat it now. However, If we think about the gain from the future state, we observe that later we can achieve more gain if we do not intervene and previously assigned resources to $c_{id} = A$ inaccurately. Hence, it is more beneficial to allocate resources to $c_{id} = B$ and apply the intervention now since we might lose current gain later. Thus, using the adjusted gain to decide when to intervene could enhance the performance of PrPM.

Selecting the best or most profitable case can be judged efficiently based on the adjusted gain to maximize the total gain. However, triggering interventions

Table 1. An example of defining gain.

c_{id}	$c\_gain(c_{id}, T_{i=1})$	$f\_gain(c_{id}, T_{i=1})$	opp_{cost}	Adju	Decision
A	7	12	5	3	Wait
B	5	1	-4	9	Treat
C	3	3	0	3	Neutral

as often as we want and immediately is impossible since resources are bounded in practice, which is the last aspect we need to consider.

Resource Allocator. Monitoring resources and assigning them to cases that need intervention is critical. The resource allocator checks the availability of resources. Once the most profitable case is selected and a free resource is available, we assign that resource to the selected case and block it for a certain time, i.e., *treatment duration* (T_{dur}). The number of available resources and the time required to perform the intervention could be identified via domain knowledge [18].

5   Evaluation

To verify the effectiveness and relevance of the proposed method, we experimentally investigate whether we can learn when to trigger an intervention to maximize the total gain at the run time, considering the tradeoff between intervening now or later. We compare our results to baselines that consider either predictive models without quantifying the prediction uncertainty [6,14] or only the current case's execution state scores [2,18] as state-of-the-art baselines by addressing the following research questions:

RQ1. To what extent does taking into account the current uncertainty prediction enhance the total gain?

RQ2. To what extent does taking into account the derivative of the uncertainty prediction and the adjusted gain enhance the total gain?

5.1   Dataset

We use a real-life event log, named *BPIC2017*[1], publicly available from the 4TU.ResearchData. The log describes the execution of a loan origination process, and we choose this log for several reasons. First, it contains a clear notion for outcome definition and interventions utilized to show our method's efficiency. Second, this log contains 31,413 applications and 1,202,267 events, which is large enough and frequently used for predictive and prescriptive methods.

---

[1] https://doi.org/10.4121/uuid:5f3067df-f10b-45da-b98b-86ae4c7a310b.

The *BPIC2017* log is characterized by various case and event attributes, and we include all of them in our experiments. Additionally, we extracted other essential critical features in our work, such as the number of sent offers, event number, and additional temporal features.

We used all original and extracted attributes as input for ensemble and causal models in the preprocessing step. Then we defined case outcomes based on the end state of cases, i.e., *"A_Pending"* state means positive outcome and *"A_canceled or A_Denied"* means negative outcome. After that, we defined the intervention that positively impacts the negative outcome as sending a second offer (or *"Creat_Offer"* activity) to all customers who received only one offer. Accordingly, we denoted cases with $T = 1$ based on the number of offers sent to each, i.e., cases that receive only one offer. Then, we extracted length prefixes no more than the 90th percentile for each case to avoid bias from lengthy cases. Finally, we used an *aggregate encoding* to encode the extracted prefixes.

5.2   Experimental Setup

The experiments show our method's effectiveness during operation time with two main objectives: deciding when to intervene, either now or later, and selecting the most profitable case among all candidates to maximize the *total gain*.

We adapted an ensemble model based on a *Gradient Boosting Decision Tree (GBDT)* method to estimate the avg_{pred} and $total_{uncer}$. In particular, we used *Catboost* [17], an open-source GBDT library with several tools to quantify the prediction uncertainty and automatically handle categorical features.

Catboost is trained with a *negative log-likelihood* loss and *Langevin optimization* [21] to ensure global conversion instead of a local optimum and generate an ensemble of several independent GBDT. Additionally, we used the following parameters during training: ensemble size of 50, a learning rate of 0.05, a subsample of 0.82, and a max tree depth of 12.

Catboost returns a probability distribution over the case outcomes. This distribution is based on a given model version, i.e., on the seed used to initialize the model parameters before training. So we train the same model using different seed initialization and evaluating these models on the same input to obtain the avg_{pred} and its $total_{uncer}$. We use an ORF algorithm implemented in the *EconMl*[2] to train a causal model to estimate the CATE.

We follow the machine learning workflow to train both ensemble and causal models. We temporally split the data into training (60%), validation (20%), and testing (20%) sets. Training and validation sets are used to train and tune model parameters, while the testing set is used to evaluate the model's performance.

During the testing time, we follow the configurations shown in Table 2. First, ongoing cases are filtered into candidates. We filter cases based on the estimated probability of avg_{pred} >0.5 to ensure that cases are highly probable to end negatively and $CATE$ >0 to guarantee that intervention has a positive impact.

---

[2] https://github.com/microsoft/EconML.

Table 2. Parameter settings of the introduced method

avg_{proba}	$CATE$	$total_{uncer}$	$\Delta total_{uncer}$	#Resources	c_{uout}	c_{T1}	T_{dur} (sec)
>0.5	>0	< 0.25, 0.5, 0.75	< 0, −0.5, −0.1, −0.15, −0.2, −0.25, −0.3	1, 2, ...10	20	1	Fixed = 60 Normal ∈ {1, 60} Exponential ∈ {1, 60}

Additionally, we filter cases using the $total_{uncer}$ to see how sure the predictive model is with the predictions. The estimated $total_{uncer}$ ranges from 0, where the model is certain, to 1, which is entirely uncertain. We experiment with three thresholds (see Table 2) when considering only the current state of cases.

On the other hand, we use the opp_{coast} or adj_{gain} when we consider both current and future states of ongoing cases. At the same time, with and without a derivative of the total uncertainty ($\Delta total_{uncer}$), representing the difference between c_total_{uncer} and f_total_{uncer}. Accordingly, we filter cases based on the c_avg_{pred} and c_CATE and then select the case with the highest adj_{gain} with and without $\Delta total_{uncer}$ when it is below 0 or negative values (see Table 2), which means the predictive model becomes more uncertain in the future state.

5.3 Results

We show results here based on one T_{dur} distribution, i.e., fixed. Since the normal distribution achieves similar results to the fixed in terms of the total gain and is higher than the exponential distribution as the variability is higher in the latter. Also, we observed that the behavior for different thresholds for $total_{uncer}$ and $\Delta total_{uncer}$ is the same, i.e. when we lower the threshold, fewer cases are treated with no substantial effect on the total gain per case. Hence, we present results using one threshold, i.e., the higher value for $total_{uncer} < 0.75$ and $\Delta total_{uncer} < 0$. However, The full results of experimenting with all thresholds and T_{dur} distributions are available in supplementary material[3].

To discuss RQ1, we consider only the current state scores of ongoing cases. Then, examine how the total gain evolves when adding the current prediction uncertainty (c_total_{uncer}) to the filtering step. Besides the estimated prediction score (c_avg_{proba}) and the causal effect (c_CATE). We compare this to baselines shown in [3,18], where only the avg_{proba} and $CATE$ are set, see Fig. 3a.

The results in Fig. 3a explain that adding the current prediction uncertainty improves the total gain when the available resources exceed 80%. Also, the baseline treats more cases with less total gain (as shown in the supplementary (see Footnote 3)). In contrast, adding the current prediction uncertainty allows efficient allocation of resources since fewer cases are treated but with a higher total gain than the baseline. Because considering the current prediction uncertainty does not trigger the intervention until the predictive model becomes more sure about its predictions and achieves higher total gain.

---

[3] https://zenodo.org/record/6381445#.YjwaFfexWuA.

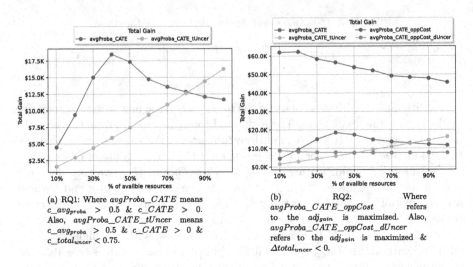

(a) RQ1: Where $avgProba\_CATE$ means $c\_avg_{proba} > 0.5$ & $c\_CATE > 0$. Also, $avgProba\_CATE\_tUncer$ means $c\_avg_{proba} > 0.5$ & $c\_CATE > 0$ & $c\_total_{uncer} < 0.75$.

(b) RQ2: Where $avgProba\_CATE\_oppCost$ refers to the adj_{gain} is maximized. Also, $avgProba\_CATE\_oppCost\_dUncer$ refers to the adj_{gain} is maximized & $\Delta total_{uncer} < 0$.

Fig. 3. The total gain progress under the same number of available resources.

Turning to discuss RQ2, we consider two states' scores of ongoing cases, i.e., current and future, to estimate the opp_{cost} and the adj_{gain} and their impacts on the total gain. Hence, we decide when to intervene and allocate the available resources to ongoing cases based on two things: first, cases with the maximum adj_{gain} only, and second, cases with the maximum adj_{gain} and the $\Delta total_{uncer} < 0$. We compare this method to the baseline in RQ1 and consider the current prediction uncertainty only, see Fig. 3b.

The results in Fig. 3b show that the total gain is highly affected by the trade off of current and future state scores under the same quantity of consumed resources. The proposed method achieves a significant improvement in terms of the total gain and outperforms the baselines (cf. Fig 3a). We found that considering the adj_{gain} as a decision function to determine when to intervene is more efficient than considering it with the $\Delta total_{uncer}$. Still, considering the derivative of the prediction uncertainty could achieve a higher gain when the available resources are below 10%. Because when the number of available resources increases, more cases are treated without ensuring that the $total_{uncer}$ is sufficient.

The proposed PrPM method results in Fig. 3 demonstrate a higher total gain than the baselines. Accordingly, estimating the prediction uncertainty of the underlying predictive model and analyzing the tradeoff between triggering an intervention now versus later based on the adjusted gain and the opportunity cost can optimize the performance of PrPM methods, hence, business processes.

5.4 Threats to Validity

Our method's evaluation has an external validity threat (lack of generalizability) because of its dependence on only one dataset. Accordingly, the evaluation is vague and needs more experiments using other logs to be followed up.

We assume that $CATE$ is accurate and will reduce the probability of negative outcomes. Also, the intervention will be triggered only once for each ongoing case. Accordingly, there is a threat to ecological reality where cases may be treated more than once via multiple interventions. Moreover, the $CATE$ may not represent the natural causal effect because of unobserved confounders.

The experiments' setup used one feature encoding technique and did not discuss the selection bias of the causal model due to variables that affect both the outcome and the intervention. Utilizing other encoding and selection bias techniques is a hint for future work to improve the performance of PrPM methods.

6 Conclusion and Future Work

We presented a prescriptive monitoring method to determine if and when an intervention should be triggered on ongoing cases to maximize a gain function. The method leverages an ensemble model to estimate the probability of negative case outcomes and the associated prediction uncertainty. They are combined with a causal model to assess the effect of an intervention on the case outcome. These estimates and a filtering and ranking method are embedded in a resource allocator. It assigns resources to perform interventions to maximize total gain, considering the tradeoff between intervening now or later. An initial evaluation shows that taking into account the possible future states of ongoing cases and the level of prediction uncertainty leads to a higher gain than baselines that rely only on the current state of each case.

The proposed prescriptive method does not include constraints on when an intervention may be triggered. In practice, it is often not possible to trigger an intervention at any point in a case. For example, sending a second loan offer to a customer who has already accepted an offer at another bank is counterproductive. A direction for future work is to extend the approach with temporal rules on the interventions and consider these rules in the intervention policy.

The proposed method is limited to handling one type of intervention. Also, an intervention is applied at most once to a case. Lifting these restrictions is another direction for future work.

Reproducibility. The implementation and source code of the method, together with instructions to reproduce the evaluation, can be found at: https://github.com/mshoush/prescriptive-monitoring-uncertainty.

References

1. Abdar, M., et al.: A review of uncertainty quantification in deep learning: techniques, applications and challenges. Inf. Fusion **76**, 243–297 (2021)

2. Bozorgi, Z.D., Teinemaa, I., Dumas, M., La Rosa, M.: Prescriptive process monitoring for cost-aware cycle time reduction. In: ICPM (2021)
3. Bozorgi, Z.D., Teinemaa, I., Dumas, M., La Rosa, M., Polyvyanyy, A.: Process mining meets causal machine learning: discovering causal rules from event logs. In: ICPM, pp. 129–136. IEEE (2020)
4. Di Francescomarino, C., Dumas, M., Maggi, F.M., Teinemaa, I.: Clustering-based predictive process monitoring. IEEE Trans. Serv. Comput. **12**(6), 896–909 (2016). https://doi.org/10.1109/TSC.2016.2645153
5. Dietterich, T.G., et al.: Ensemble learning. Handb. Brain Theor. Neural Netw. **2**(1), 110–125 (2002)
6. Fahrenkrog-Petersen, S.A., et al.: Fire now, fire later: alarm-based systems for prescriptive process monitoring. Knowl. Inf. Syst. **64**, 559–587 (2022). https://doi.org/10.1007/s10115-021-01633-w
7. Gal, Y., Ghahramani, Z.: Dropout as a bayesian approximation: representing model uncertainty in deep learning. In: ICML, pp. 1050–1059. PMLR (2016)
8. Guo, R., Cheng, L., Li, J., Hahn, P.R., Liu, H.: A survey of learning causality with data: problems and methods. ACM Comput. Surv. **53**(4), 1–37 (2020)
9. Hüllermeier, E., Waegeman, W.: Aleatoric and epistemic uncertainty in machine learning: an introduction to concepts and methods. Mach. Learn. **110**(3), 457–506 (2021). https://doi.org/10.1007/s10994-021-05946-3
10. Kim, J., Comuzzi, M., Dumas, M., Maggi, F.M., Teinemaa, I.: Encoding resource experience for predictive process monitoring. Decis. Support Syst. **153**, 113669 (2022)
11. Kratsch, W., Manderscheid, J., Röglinger, M., Seyfried, J.: Machine learning in business process monitoring: a comparison of deep learning and classical approaches used for outcome prediction. Bus. Inf. Syst. Eng. **63**(3), 261–276 (2021)
12. Maggi, F.M., Di Francescomarino, C., Dumas, M., Ghidini, C.: Predictive monitoring of business processes. In: Jarke, M., et al. (eds.) CAiSE 2014. LNCS, vol. 8484, pp. 457–472. Springer, Cham (2014). https://doi.org/10.1007/978-3-319-07881-6_31
13. Malinin, A., Prokhorenkova, L., Ustimenko, A.: Uncertainty in gradient boosting via ensembles. arXiv preprint arXiv:2006.10562 (2020)
14. Metzger, A., Kley, T., Palm, A.: Triggering proactive business process adaptations via online reinforcement learning. In: Fahland, D., Ghidini, C., Becker, J., Dumas, M. (eds.) BPM 2020. LNCS, vol. 12168, pp. 273–290. Springer, Cham (2020). https://doi.org/10.1007/978-3-030-58666-9_16
15. Pauwels, S., Calders, T.: Incremental predictive process monitoring: the next activity case. In: Polyvyanyy, A., Wynn, M., Van Looy, A., Reichert, M. (eds.) BPM 2021. LNCS, vol. 12875, pp. 123–140. Springer, Cham (2021). https://doi.org/10.1007/978-3-030-85469-0_10
16. Peterson, L.E.: K-nearest neighbor. Scholarpedia **4**(2), 1883 (2009)
17. Prokhorenkova, L., Gusev, G., Vorobev, A., Dorogush, A.V., Gulin, A.: CatBoost: unbiased boosting with categorical features. In: NeurIPS 31 (2018)
18. Shoush, M., Dumas, M.: Prescriptive process monitoring under resource constraints: a causal inference approach. In: Munoz-Gama, J., Lu, X. (eds.) ICPM 2021. LNBIP, vol. 433, pp. 180–193. Springer, Cham (2022). https://doi.org/10.1007/978-3-030-98581-3_14
19. Stone, J.V.: Bayes' Rule: A Tutorial Introduction to Bayesian Analysis. Sebtel Press, Sheffield (2013)

20. Teinemaa, I., Dumas, M., Rosa, M.L., Maggi, F.M.: Outcome-oriented predictive process monitoring: review and benchmark. ACM Trans. Knowl. Disc. Data **13**(2), 1–57 (2019)
21. Ustimenko, A., Prokhorenkova, L.: SGLB: stochastic gradient Langevin boosting. In: International Conference on Machine Learning, pp. 10487–10496. PMLR (2021)
22. Verenich, I., Dumas, M., Rosa, M.L., Maggi, F.M., Teinemaa, I.: Survey and cross-benchmark comparison of remaining time prediction methods in business process monitoring. ACM Trans. Intell. Syst. Technol. **10**(4), 1–34 (2019)
23. Weinzierl, S., Dunzer, S., Zilker, S., Matzner, M.: Prescriptive business process monitoring for recommending next best actions. In: Fahland, D., Ghidini, C., Becker, J., Dumas, M. (eds.) BPM 2020. LNBIP, vol. 392, pp. 193–209. Springer, Cham (2020). https://doi.org/10.1007/978-3-030-58638-6_12
24. Weytjens, H., De Weerdt, J.: Learning uncertainty with artificial neural networks for improved remaining time prediction of business processes. In: Polyvyanyy, A., Wynn, M.T., Van Looy, A., Reichert, M. (eds.) BPM 2021. LNCS, vol. 12875, pp. 141–157. Springer, Cham (2021). https://doi.org/10.1007/978-3-030-85469-0_11
25. Xu, G., Duong, T.D., Li, Q., Liu, S., Wang, X.: Causality learning: a new perspective for interpretable machine learning. arXiv preprint arXiv:2006.16789 (2020)

Author Index

Printed in the United States
by Baker & Taylor Publisher Services